Darwin without Malthus

Monographs on the History and Philosophy of Biology

RICHARD BURIAN, RICHARD BURKHARDT, JR.,
RICHARD LEWONTIN, JOHN MAYNARD SMITH

EDITORS

The Cuvier-Geoffrey Debate: French Biology in the Decades Before Darwin
TOBY A. APPEL

Controlling Life: Jacques Loeb and the Engineering Ideal in Biology
PHILIP J. PAULY

Beyond the Gene: Cytoplasmic Inheritance and the Struggle for Authority in Genetics
JAN SAPP

The Heritage of Experimental Embryology: Hans Spemann and the Organizer
VIKTOR HAMBURGER

The Evolutionary Dynamics of Complex Systems: A Study in Biosocial Complexity
C. DYKE

The Wellborn Science: Eugenics in Germany, France, Brazil, and Russia
Edited by MARK B. ADAMS

Darwin without Malthus: The Struggle for Existence in Russian Evolutionary Thought
DANIEL P. TODES

Darwin without Malthus

The Struggle for Existence in Russian Evolutionary Thought

DANIEL P. TODES

New York   Oxford
OXFORD UNIVERSITY PRESS
1989

Oxford University Press

Oxford   New York   Toronto
Delhi   Bombay   Calcutta   Madras   Karachi
Petaling Jaya   Singapore   Hong Kong   Tokyo
Nairobi   Dar es Salaam   Cape Town
Melbourne   Auckland

and associated companies in
Berlin   Ibadan

Copyright © 1989 by Oxford University Press, Inc.

Published by Oxford University Press, Inc.,
200 Madison Avenue, New York, New York 10016

Oxford is a registered trademark of Oxford Univeristy Press

All rights reserved. No part of this publication may be reproduced,
stored in a retrieval system, or transmitted, in any form or by any means,
electronic, mechanical photocopying, recording, or otherwise,
without the prior permission of Oxford University Press.

Library of Congress Cataloging-in-Publication Data
Todes, Daniel Philip.
Darwin without Malthus : the struggle for existence in Russian
evolutionary thought / Daniel P. Todes.
p. cm.— (Monographs on the history and philosophy of
biology)
Includes index.
ISBN 0-19-505830-5
1. Natural selection—History.   2. Biology—Soviet Union—History.
3. Biology—Soviet Union—Philosophy—History.   4. Darwin, Charles,
1809–1882.   5. Malthus, T. R. (Thomas Robert), 1766–1834.
I. Title   II. Series.
QH375.T63   1989   88-34301
575.01′62′09—dc19   CIP

2 4 6 8 9 7 5 3 1

Printed in the United States of America

For my parents

Acknowledgments

Writing this book has been a gratifyingly cooperative experience. It is the product of a continual dialogue with Mark B. Adams, my mentor in the history of biology and Russian science. The ideas, organization, and prose in this volume were shaped in large part by his incisive critiques of successive drafts and often emerged in the course of our innumerable conversations about this subject. His moral support has been indispensable and his intellectual example inspiring.

This volume has also benefitted from the friendly criticism of other colleagues, including audiences at the Darwin in Russia Symposium at the University of Pennsylvania (1983); the history of science colloquia at the University of Pennsylvania (1984), Princeton University (1985), and The Johns Hopkins University (1985); and the annual meeting of the History of Science Society (1986). My thanks also to David Englestein, Anne Marie Moulin, and Arthur Silverstein for their perceptive comments about drafts of particular chapters; to Linda Berg, Howard Gruber, V. A. Markin, Ernst Mayr, Edgar Melton, and Alfred Rieber for their helpful responses to specific inquiries; and to Alan Derickson for suggesting this volume's title.

Dolores Sawicki performed numerous research tasks, reviewed the manuscript repeatedly with a much more careful eye for detail than my own, and prepared it for the press. I am also indebted to the staffs of the Hoover Institution on War, Revolution and Peace, the Library of Congress; the interlibrary loan offices of the Helsinki University Library; and the Welch Medical Library of The Johns Hopkins University. My thanks to the Hoover Institution for permission to cite archival material and to the History of Science Society for permission to reproduce passages of my article "Darwin's Malthusian Metaphor and Russian Evolutionary Thought, 1859–1917," *ISIS* 78 (1987).

I am very grateful to Gert H. Brieger, Director of our program at the Institute of the History of Medicine at The Johns Hopkins University, whose support and confidence made completion of this volume possible. It was my great fortune to undergo the ordeal of writing a first book simultaneously with my good friend Marc Levine, and to be able to draw on his empathy, wisdom, and good humor. Shelby Morgan made the leap of faith that all this lunacy would amount to something and sustained me with her strength, friendship, and love. Sarah helped, too, in her way.

Contents

Introduction, 3
1. Darwin's Metaphor and His Russian Audience, 7
2. Malthus, Darwin, and Russian Social Thought, 24
3. Beketov, Botany, and the Harmony of Nature, 45
4. Korzhinskii, the Steppe, and the Theory of Heterogenesis, 62
5. Mechnikov, Darwinism, and the Phagocytic Theory, 82
6. Kessler and Russia's Mutual Aid Tradition, 104
7. Kropotkin's Theory of Mutual Aid, 123
8. Severtsov, Timiriazev, and the Classical Tradition, 143

Conclusion, 166

Notes, 173

Name Index, 213

Subject Index, 217

Darwin without Malthus

Introduction

The relationship between scientific thought and the broader context in which it develops is a tantalizing, almost irresistible subject. It is central to the great drama of the history of science—the effort of real human beings to understand an infinitely complex reality. Scientists must rely on some theoretical framework to organize the gathering and interpretation of information, and this framework inevitably has some relationship, not only to nature itself, but also to the circumstances in which scientists live, work, and think.

This book addresses one episode in this drama by examining the crosscultural transmission of a metaphor in scientific thought. Specifically, it explores the fate of an expression—the "struggle for existence"—utilized by a member of one culture, the Englishman Charles Darwin, to explain his selection theory to members of a quite different culture, the intellectuals of tsarist Russia.[1]

My central contention is that the critical reaction of Russian intellectuals to this metaphor and its conceptual implications had substantial consequences for Russian evolutionary thought. For Darwin and other leading British evolutionists, the expression "struggle for existence" appealed to common sense, and its Malthusian associations posed no problem. For Russian intellectuals, however, it was at best imprecise and confusing; at worst, and this was much more common, it was fallacious and offensive. They reacted negatively to what they perceived as a transparent introduction of Malthusianism—or, for some, simply the British enthusiasm for competition—into evolutionary theory. Very few thought this flaw justified total rejection of Darwin's ideas. A more common reaction was to break down the struggle for existence into its component parts, to explore their relationship and relative importance in nature, and to conclude that Darwin had greatly exaggerated the role of the two parts most closely identified with Malthus: that is, of overpopulation as the generator of conflict and of intraspecific competition as its result. This response defined a general direction of inquiry. Yet each individual scientist— influenced by his own biological material, disciplinary training, ideological orientation, and various biographical factors—took a different path.

I suggest that this response followed from the material circumstances of Russian life—from its class structure and political traditions, and from the characteristics of its land and climate. These gave rise to a mutually reinforcing combination of "anti-Malthusian" and "non-Malthusian" sentiments. By anti-Malthusian I refer to the negative reaction to Malthus's political doctrine that characterized Russian intellectuals across the political spectrum. By non-Malthusian I refer to the lack of commonsensical resonance between the experiences of Russians living on a vast continental plain and Malthus's vision of a natural world in which organisms were pressed into mutual conflict by population pressures. Anti-Malthusianism so permeated Russians' comments about Darwin's theory that I am confident the reader will soon grant me this point. The evidence for non-Malthusianism is less decisive, I think, because it was expressed more subtly; it often appeared as a simple description of nature ("nature is but thinly populated and features the struggle of organisms against the elements") or within an explicitly anti-Malthusian assertion ("only a naturalist blinded by Malthusianism would contend that nature is overpopulated and that conflicts within it transpire primarily between organism and organism").

As one would expect, such a pervasive characteristic of Russian evolutionary thought has been noticed previously. The more cosmopolitan Russian intellectuals of the nineteenth century, including P. A. Kropotkin and K. A. Timiriazev, themselves remarked on the distinctive anti-Malthusianism of their countrymen. From the 1930s through the 1950s, Lysenkoist historians, seeking to portray themselves as the rightful heirs to a proud national tradition, applauded the perspicacity of earlier Russian evolutionists who had identified and corrected Darwin's "Malthusian error." In the 1970s K. M. Zavadskii commented on the Russian preoccupation with "the problem of the struggle for existence," and Ia. M. Gall explored the views of Russian botanists in a rich monograph entitled *The Struggle for Existence as a Factor in Evolution*. In the West, James Allen Rogers has written several illuminating essays about the criticism of Darwin's Malthusianism by Russia's social thinkers, and Alexander Vucinich has also noted Russian skepticism about the struggle for existence. As this present volume was nearing completion, Francesco M. Scudo and Michele Acanfora published a fine essay on Darwin in Russia in which they discussed this skepticism briefly but perceptively.[2]

This book is organized as follows: Chapter 1 introduces Darwin's concept of the struggle for existence, notes the unproblematic association of this metaphor with Malthus by Darwin and other leading British evolutionists, and describes the audience that awaited it in Russia. Chapter 2 explores the overwhelmingly negative reaction to Malthus by Russian social thinkers, documents their frequent association of Darwin's theory with Malthus's political doctrine, and examines the different ways in which they analyzed the Darwin-Malthus connection. The core of this volume, chapters 3 through 8, is devoted to the scientific work of a number of important Russian scientists, representing three generations and a range of biological disciplines and political orientations. As these figures are largely unfamiliar to Western readers, I begin each chapter with a biographical sketch before discussing the scientist's reaction to Darwin's metaphor both in general formulations and in approaches to specific scientific issues. A. N. Beketov (1825–1902) was Russia's premier botanist of the nineteenth century, an evolutionist before 1859, a liberal, and a religious man. S. I. Korzhinskii (1861–1900) was the controversial botanical geographer and political conservative who developed the "theory of heterogenesis," a mutation theory published two years before Hugo de Vries's *Die Mutation-*

stheorie. I. I. Mechnikov (1845-1916) was a zoologist and pathologist, a liberal and an atheist, and the Nobel Prize-winning author of the phagocytic theory of inflammation. K. F. Kessler (1815-1881) was the eminent and politically moderate ichthyologist who became the seminal theorist of Russia's mutual aid tradition. P. A. Kropotkin (1842-1921) was the geologist, geographer, and anarchist with whom Westerners usually associate this tradition. Finally, I turn to two classical Darwinians who constitute illustrative exceptions to the rule: the pioneer-naturalist N. A. Severtsov (1827-1885) and "Darwin's Russian Bulldog," plant physiologist K. A. Timiriazev (1843-1920). These seven individuals, together with the many others who are treated less extensively in this volume—including V. M. Bekhterev, M. N. Bogdanov, A. F. Brandt, N. Ia. Danilevskii, V. V. Dokuchaev, V. O. Kovalevskii, N. F. Levakovskii, M. A. Menzbir, N. D. Nozhin, and I. S. Poliakov—hardly exhaust the list of relevant thinkers, but they do include almost every major figure and provide, I think, a fair representation of Russian views.[3]

Having described what this book is, let me make clear what it is not. It is not a comprehensive history of Russian responses to Darwin; that would require examination of a multitude of institutions, issues, and political and philosophical tendencies that the reader will not encounter here. Nor does my argument rest on any particular view of the actual relationship between Malthus's ideas and Darwin's creative path to the selection theory. Russians could not respond to that relationship (whatever it was) but only to Darwin's quite explicit association of his views with Malthus's and to their own identification of a connection between the two thinkers.

A few final introductory comments are in order. First, I have attempted to present each scientist's views on the struggle for existence essentially as he himself did. Philosophical assumptions, anthropomorphisms, and political implications so permeate these arguments that their flow would be entirely disrupted if I were to dwell on each one. I often rely, therefore, on the perspicacity of the reader. Second, although I have organized this book largely around individuals, I can offer only general suggestions regarding the process by which each drew his conclusions. As we know from numerous studies, the relationship between this process of scientific creativity, on the one hand, and a scientist's self-presentation in publications and reminiscences, on the other, is problematic. Yet the admittedly complex individuality of each Russian's thinking conformed to a general pattern. My goal is to expose that pattern, to suggest two general sources of it, and to explore the intellectual consequences of Russians' attempts to separate Darwin from Malthus.

CHAPTER 1

Darwin's Metaphor and His Russian Audience

Every one knows what is meant and is implied by such metaphorical expressions, and they are almost necessary for brevity.[1] (Charles Darwin, 1861)

By his use of metaphor Darwin appealed to the common knowledge of his audience about certain subjects in order to enlighten them about another, the relationships in nature that resulted in evolution. His metaphors resonated, albeit unequally, within influential and relatively broad sectors of British society, and so added poetic vision and commonsensical power to his argument. Such rhetorical authority contributes to a metaphor's cognitive function, enabling it to clarify certain points and obscure others, to encourage exploration of certain questions and distinctions, and to relegate others to relative unimportance.

This chapter introduces several subjects important to the transmission of Darwin's metaphor "struggle for existence" from Great Britain to Russia. I begin by discussing what Darwin meant by this expression. Then I examine its unproblematic association with Malthus for Darwin and other leading British evolutionists, and note the particular aspects of the "struggle for existence" that plausibly might have been identified with Malthus's influence. I end with a sketch of the audience that awaited Darwin's Malthusian metaphor in Russia.

Darwin on the Struggle for Existence

"Nothing is easier than to admit in words the truth of the universal struggle for life," wrote Charles Darwin in *On the Origin of Species by Means of Natural Selection or the Preservation of Favoured Races in the Struggle for Life*, "or more difficult than constantly to bear this conclusion in mind."[2]

To ease this difficulty Darwin devoted the third, pivotal chapter of the *Origin* entirely to the struggle for existence. The two previous chapters had prepared the analogy between artificial and natural selection. Chapter 1, "Variation Under Domestication," concerned the wide range of variations found in domesticated animals and the selective breeding of varieties among them. Chapter 2, "Variation Under Nature," established the great number of individual variations in nature. The reader was thus prepared for an explanation of nature's counterpart to artificial selection, which Darwin presented in Chapter 3, "Struggle for Existence," and Chapter 4, "Natural Selection."

Introducing his concept of the struggle for existence Darwin explained:

> I use the term Struggle for Existence in a large and metaphorical sense, including dependence of one being on another, and including (which is more important) not only the life of the individual, but success in leaving progeny. Two canine animals in a time of dearth, may be truly said to struggle with each other which shall get food and live. But a plant on the edge of a desert is said to struggle for life against the drought, though more properly it should be said to be dependent on the moisture. A plant which annually produces a thousand seeds, of which on an average only one comes to maturity, may be more truly said to struggle with the plants of the same and other kinds which already clothe the ground. The missletoe is dependent on the apple and a few other trees, but can only in a far-fetched sense be said to struggle with these trees, for if too many of these parasites grow on the same tree, it will languish and die. But several seedling missletoes, growing close together on the same branch, may more truly be said to struggle with each other. As the missletoe is disseminated by birds, its existence depends on birds; and it may metaphorically be said to struggle with other fruit-bearing plants, in order to tempt birds to devour and thus disseminate its seeds rather than those of other plants. In these several senses, which pass into each other, I use for convenience sake the general term of struggle for existence.[3]

The "struggle for existence," then, was a metaphor for a number of different relationships among organisms, and between the organism and physical conditions, that resulted in the survival and reproductive success of some individuals, and the death and less frequent reproduction of others.

This struggle for existence followed necessarily from the rate at which organisms reproduced: "As more individuals are produced than can possibly survive," Darwin explained, "there must in every case be a struggle for existence, either one individual with another of the same species, or with the individuals of distinct species, or with the physical conditions of life. It is the doctrine of Malthus applied with manifold force to the whole animal and vegetable kingdoms."[4]

Darwin acknowledged that naturalists knew little about the actual dynamics of the struggle for existence in any specific case. Try to imagine how to give one form an advantage over another, he suggested. "Probably in no single instance should we know what to do, so as to succeed. It will convince us of our ignorance on the mutual relations of all organic beings; a conviction as necessary, as it seems to be difficult to acquire."[5]

Such ignorance did not, however, compromise a fundamental truth: the tangle of different relations in nature conferred an overall advantage on some individuals. This summary advantage provided the material for natural selection:

> If there be, owing to the high geometrical powers of increase of each species, at some age, season, or year, a severe struggle for life, and this certainly cannot be disputed; then, considering the infinite complexity of the relations of all organic beings to each other and to their conditions of existence, causing an infinite diversity in structure, constitution, and habits, to be advantageous to them, I think it would be a most extraordinary fact if no variation ever had occurred useful to each

being's own welfare, in the same way as so many variations have occurred useful to man. But if variations useful to any organic being do occur, assuredly individuals thus characterised will have the best chance of being preserved in the struggle for life; and from the strong principle of inheritance they will tend to produce offspring similarly characterised. This principle of preservation, I have called, for the sake of brevity, Natural Selection.[6]

For Darwin these inextricably linked processes of the struggle for existence and natural selection were principally responsible for the divergence of characters and the emergence of new species.[7]

Although he was primarily interested in the struggle for existence as a totality with a single end result (the advantage of one form over another), Darwin did develop several points about its actual dynamics. First, the central aspect of the struggle for existence was not the relationship between an organism and physical conditions, but rather that among organisms themselves. As one traveled from a damp to a dry region, for example, some species grew rarer and finally disappeared. One might attribute this to the fact that these species were unable to survive in a dry climate, "but this," Darwin insisted, "is a very false view." Physical conditions acted "in main part indirectly by favoring other species." This truth was readily apparent from "the prodigious number of plants in our gardens which can perfectly well endure our climate, but which never become naturalised, for they cannot compete with our native plants, nor resist destruction by our native animals."[8]

Darwin's insistence upon the centrality of organism-organism relations was related to his perception that nature was packed to capacity with living beings. One form could increase its numbers only at the expense of another. "The face of Nature," he wrote, "may be compared to a yielding surface, with ten thousand sharp wedges packed close together and driven inwards by incessant blows, sometimes one wedge being struck, and then another with greater force."[9]

Yet he acknowledged that organisms were not everywhere packed wedgelike together and that the dynamics of the struggle for existence differed therefore in different physical settings:

> Even when climate, for instance extreme cold, acts directly, it will be the least vigorous, or those which have got least food through the advancing winter, which will suffer most . . . . Each species, even where it most abounds, is constantly suffering enormous destruction at some period of its life, from enemies or from competitors for the same place and food; and if these enemies or competitors be in the least degree favoured by any slight change of climate, they will increase in numbers, and, *as each area is already fully stocked with inhabitants*, the other species will decrease. When we travel southward and see a species decreasing in numbers, we may feel sure that the cause lies quite as much in other species being favoured, as in this one being hurt. So it is when we travel northward, but in a somewhat lesser degree, *for the number of species of all kinds, and therefore of competitors, decreases northwards*; hence in going northward, or in ascending a mountain, we far oftener meet with stunted forms, due to the directly injurious action of climate, than we do in proceeding southwards or in descending a mountain. When we reach the Arctic regions, or snowcapped summits, or absolute deserts, *the struggle for life is almost exclusively with the elements* [my emphasis].[10]

Some regions, then, were not teeming with life; there, not the wedging of organism against organism, but rather the metaphorical struggle of the organism against the elements, assumed primary importance. In Chapter 6, "Difficulties of Theory,"

Darwin again specified that natural selection acted principally through competition between life forms "in each well-stocked country."[11]

This qualification apparently concerned him little, however, for nature in the *Origin* remained an essentially superfecund and plenitudinous "entangled bank."[12] This was one in a series of images, including that of a "hothouse" and an "entangled jungle," that Darwin had originally employed in reference to the particular features of tropical nature. Indeed, the attempt to describe the natural setting in Bahia, Brazil, to his countrymen familiar only with the temperate zone produced some of Darwin's most poetic and evocative prose:

> Such are the elements of the scenery, but it is a hopeless attempt to paint the general effect. Learned naturalists describe these scenes of the tropics by naming a multitude of objects, and mentioning some characteristic feature of each. To a learned traveller this possibly may communicate some definite ideas: but who else from seeing a plant in an herbarium can imagine its appearance when growing in its native soil? Who from seeing choice plants in a hothouse, can magnify some into the dimensions of forest trees, and crowd others into an entangled jungle? Who when examining in the cabinet of the entomologist the gay exotic butterflies, and singular cicadas, will associate with these lifeless objects, the ceaseless harsh music of the latter, and the lazy flight of the former,—the sure accompaniments of the still, glowing noonday of the tropics? It is when the sun has attained its greatest height, that such scenes should be viewed: then the dense splendid foliage of the mango hides the ground with its darkest shade, whilst the upper branches are rendered from the profusion of light of the most brilliant green. In the temperate zones the case is different—the vegetation there is not so dark or so rich, and hence the rays of the declining sun, tinged of a red, purple, or bright yellow colour, add most to the beauties of those climes. When quietly walking along the shady pathways, and admiring each successive view, I wished to find language to express my ideas. Epithet after epithet was found too weak to convey to those who have not visited the intertropical regions, the sense of delight which the mind experiences. I have said that the plants in a hothouse fail to communicate a just idea of the vegetation, yet I must recur to it. The land is one great wild untidy, luxuriant hothouse, made by Nature for herself, but taken possession of by man, who has studded it with gay houses and formal gardens. How great would be the desire in every admirer of nature to behold, if such were possible, the scenery of another planet! yet to every person in Europe, it may be truly said, that at the distance of only a few degrees from his native soil, the glories of another world are opened to him. In my last walk I stopped again and again to gaze on these beauties, and endeavoured to fix in my mind for ever, an impression which at the time I knew sooner or later must fail. The form of the orange-tree, the cocoanut, the palm, the mango, the tree-fern, the banana, will remain clear and separate; but the thousand beauties which unite these into one perfect scene must fade away; yet they will leave, like a tale heard in childhood, a picture full of indistinct, but beautiful figures.[13]

The codiscoverer of natural selection, Alfred Russel Wallace, later discussed the coincidence between the specific attributes of this "entangled jungle," on the one hand, and the emphasis on organism-organism relations in the selection theory, on the other.[14]

A second critical point about the dynamics of the struggle for existence was that it was most severe among like forms. "The struggle almost invariably will be most severe between the individuals of the same species, for they frequent the same districts, require the same food, and are exposed to the same dangers," Darwin wrote.[15] "Almost equally severe" was the struggle among varieties of the same species; struggle between species of the same genus was more intense than between species of distinct genera.[16]

Throughout the *Origin* Darwin used the words "struggle" and "competition" interchangeably. He did not address the relative importance of indirect competition and direct conflict among organisms, but gave the former greater weight in his book. His unification of all aspects of natural relations within the metaphor "struggle for existence," and in such phrases as "the great battle for life" and the "war of nature," contributed a certain rhetorical power to his argument, at least for a British audience. Consider his use of battle imagery in the final passage of his chapter on the struggle for existence: "When we reflect on this struggle, we may console ourselves with the full belief, that the war of nature is not incessant, that no fear is felt, that death is generally prompt, and that the vigorous, the healthy and the happy survive and multiply."[17] Here Darwin sacrificed precision for eloquence: in the metaphorical sense, his struggle for existence *is* incessant (although it varies in severity) and death is *not* generally prompt.[18]

In the *Origin* the struggle for existence transpired almost entirely among individuals, but Darwin did invoke group selection to explain sterile forms among ants: "If such insects had been social, and it had been profitable to the community that a number should have been annually born capable of work, but incapable of procreation," he argued, "I can see no very great difficulty in this being effected by natural selection."[19] In *Descent of Man* (1871) his effort to establish the continuity between animals and humans led him to address more fully the evolutionary origins of sociality. He observed that "animals of many kinds are social" and discussed cooperation among insects, rabbits, sheep, monkeys, horses, cows, wolves, and birds. This he attributed in part to "mere habit," but chiefly to natural selection:

> For with those animals which were benefited by living in close association, the individuals which took the greatest pleasure in society would best escape various dangers; whilst those that cared least for their comrades and lived solitary would perish in greater numbers. With respect to the origin of the parental and filial affections, which apparently lie at the basis of the social affections, it is hopeless to speculate; but we may infer that they have been to a large extent gained through natural selection.[20]

In *Descent of Man* cooperation often conferred an advantage on individuals (as in the above citation) and on a group as a whole: "Those communities, which included the greatest number of the most sympathetic members, would flourish best and rear the greatest number of offspring."[21] Darwin emphasized, however, that such cooperation extended only to "associated groups," not to species per se, and that natural selection could not alter one species for the exclusive benefit of another.

Darwin never commented directly on the relationship between cooperation and the struggle for existence among individuals. It may well not have occurred to him that some might identify a tension between the combat and dependency relations that he included within his metaphor. It was left to the reader to infer that, for Darwin, cooperating forms were still engaged in a metaphorical struggle for existence among themselves.

Although his ideas about evolution had always been inseparable from his thoughts about the genesis and nature of man, Darwin confined himself in the *Origin* to the suggestion that "light will be thrown on the origin of man and his history."[22] In *Descent of Man* he explained that the same factors responsible for evolution by natural selection among lower organisms—variation, heredity, and the struggle for existence—obtained also among humans:

> Man is liable to numerous, slight, and diversified variations, which are induced by the same general causes, are governed and transmitted in accordance with the same general laws, as in the lower animals. Man tends to multiply at so rapid a rate that his offspring are necessarily exposed to a struggle for existence, and consequently to natural selection.[23]

Even an inattentive reader of *Descent of Man* would have encountered numerous passages in which Darwin's discussion of the struggle for existence among humans reflected his "Liberal or Radical" political views.[24] Let us consider only two that raised the broader social implications of his biological theory:

> Natural selection follows from the struggle for existence; and this from a rapid rate of increase. It is impossible not bitterly to regret, but whether wisely is another question, the rate at which man tends to increase; for this leads in barbarous tribes to infanticide and many other evils, and in civilised nations to abject poverty, celibacy, and to the late marriages of the prudent. But as man suffers from the same physical evils with the lower animals, he has no right to expect an immunity from the evils consequent on the struggle for existence. Had he not been subjected to natural selection, assuredly he would never have attained to the rank of manhood. When we see in many parts of the world enormous areas of the most fertile land peopled by a few wandering savages, but which are capable of supporting numerous happy homes, it might be argued that the struggle for existence had not been sufficiently severe to force man upwards to his highest standard. Judging from all that we know of man and the lower animals, there has always been sufficient variability in the intellectual and moral faculties, for their steady advancement through natural selection. No doubt such advancement demands many favourable concurrent circumstances; but it may well be doubted whether the most favourable would have sufficed, had not the rate of increase been rapid, and the consequent struggle for existence severe to an extreme degree.[25]

And in the concluding chapter:

> The advancement of the welfare of mankind is a most intricate problem; all ought to refrain from marriage who cannot avoid abject poverty for their children; for poverty is not only a great evil, but tends to its own increase by leading to recklessness in marriage. On the other hand, as Mr.Galton has remarked, if the prudent avoid marriage, whilst the reckless marry, the inferior members will tend to supplant the better members of society. Man, like every other animal, has no doubt advanced to his present high condition through a struggle for existence consequent on his rapid multiplication; and if he is to advance still higher he must remain subject to a severe struggle. Otherwise he would soon sink into indolence, and the more highly-gifted men would not be more successful in the battle of life than the less gifted. Hence our natural rate of increase, though leading to many and obvious evils, must not be greatly diminished by any means. There should be open competition for all men; and the most able should not be prevented by laws or customs from succeeding best and rearing the largest number of offspring. Important as the struggle for existence has been and even still is, yet as far as the highest part of

man's nature is concerned there are other agencies more important. For the moral qualities are advanced, either directly or indirectly, much more through the effects of habit, the reasoning powers, instruction, religion, &c., than through natural selection; though to this latter agency the social instincts, which afforded the basis for the development of the moral sense, may be safely attributed.[26]

The point here is not to identify Darwin with any particular current of "Social Darwinism" nor, clearly, with an unalloyed Malthusian outlook. It is, rather, to illustrate the unsurprising fact that he shared the ideological outlook of his class, circle, and family. This outlook was not universal, and a reader of *Descent of Man* who did not share it—who did not agree that the ideal society provided "open competition for all men," that such competition favored "the more highly-gifted" and led to progress, that the poor should limit their numbers, and so forth—such a reader might easily identify the author's ideological preconceptions as bourgeois, Malthusian, or, perhaps, typically British.

The Malthus Connection

Twice in the *Origin* Darwin characterized his concept of the struggle for existence as "the doctrine of Malthus applied with manifold force to the whole animal and vegetable kingdoms." This was but one of several occasions on which he and his allies associated this metaphor with Malthus's doctrine. It is this common association, which has a historical reality independent of Malthus's actual role in Darwin's creative process, to which I refer by the phrase "the Malthus connection."

The Reverend Thomas Robert Malthus (1766–1834) was the product of a merchant family and the father of three children. In 1798 he published *An Essay on the Principle of Population, as It Affects the Future Improvement of Society*, a short, trenchant critique of Godwin, Condorcet, and other advocates of the Enlightenment view that humans and human society were infinitely perfectible.[27] For Malthus such optimism foundered on an inexorable natural law: "Population, when unchecked, increases in a geometrical ratio. Subsistence increases only in an arithmetical ratio." Thus, all organisms were subject to "a strong and constantly operating check on population." Among plants and animals this check took the form of "waste of seed, sickness, and premature death." Among humans it was expressed in "misery and vice." More specifically, Malthus identified "positive checks" (such as war, famine, and epidemics), which raised the mortality rate, and less significant "preventive checks" (such as postponement of marriage and childbearing), which lowered the birthrate. Both forms of "misery and vice" were ultimately attributable to the relative scarcity of provisions.[28]

Malthus built his argument on the familiar terrain that humans, like other organisms, were subject to natural law—and he illustrated his argument with many examples from the plant and animal kingdoms. When resources were plentiful, organic populations increased rapidly, only to be "repressed afterwards by want of room and nourishment" and, among animals, by beasts of prey.[29] Human tribes expanded similarly in times of plenty and sought new territories to accommodate their larger numbers. An expanding tribe easily overwhelmed a peaceful neighbor, but when it encountered one driven, like itself, by population pressures, "the contest was a struggle for existence; and they fought with a desperate courage, inspired by the reflection, that death was the punishment of defeat, and life the prize of victory."[30]

Wars, famines, epidemics, and other scourges were, then, inherent to human existence. With equal inevitability these ills fell unevenly on the population. Nature's creations were not equally perfect, Malthus observed, and "where there is any inequality of conditions . . . the distress arising from a scarcity of provisions, must fall hardest upon the least fortunate members of society."[31]

For Malthus the relentlessly imbalanced progressions constituted a powerful force for stasis in nature and society. Ridiculing Condorcet's speculations about "the organic perfectibility, or degeneration, of the race of plants and animals," Malthus observed that even breeders of domesticated animals operated within sharply defined limits. They had, for example, succeeded in breeding Leicester sheep with small heads and legs, but "I should not scruple to assert, that were the breeding to continue for ever, the head and legs of these sheep would never be so small as the head and legs of a rat."[32] Those who projected a future of harmony and prosperity for all humans ignored nature's severe strictures on change. One might write with equal authority of man's future as an ostrich:

> A writer may tell me that he thinks man will ultimately become an ostrich. I cannot properly contradict him. But before he can expect to bring any reasonable person over to his opinion, he ought to show, that the necks of mankind have been gradually elongating; that the lips have grown harder, and more prominent; that the legs and feet are daily altering their shape; and that the hair is beginning to change into stubs of feathers. And till the probability of so wonderful a conversion can be shown, it is surely lost time and lost eloquence to expatiate on the happiness of man in such a state; to describe his powers, both of running and flying; to paint him in a condition where all narrow luxuries would be condemned; where he would be employed only in collecting the necessaries of life; and where, consequently, each man's share of labour would be light, and his portion of leisure ample.[33]

Visions of social progress and reforms toward that end were, then, chimerical:

> I see no way by which man can escape from the weight of this law which pervades all animated nature. No fancied equality, no agrarian regulations in their utmost extent, could remove the pressure of it even for a single century. And it appears, therefore, to be decisive against the possible existence of a society, all the members of which, should live in ease, happiness, and comparative leisure; and feel no anxiety about providing the means of subsistence for themselves and families.[34]

For Malthus, then, rather than pursuing unnatural and unrealizable goals, nations should abolish impediments to free competition. Britain's system of parish poor laws, for example, was, like all charity, shortsighted and ultimately counterproductive. It increased population but not provisions, and so diminished the amount of food available to the industrious poor; it also encouraged dissipation at the expense of frugality and independence.

Malthus's original *Essay* was a hurried, political pamphlet. In 1803 he issued a much-revised second edition, entitled *An Essay on the Principle of Population, or a View of Its Past and Present Effects on Human Happiness with an Inquiry into our Prospects Respecting the Future Removal or Mitigation of the Evils Which It Occasions.* Here he examined in detail the various checks on population growth in different countries and emphasized one, "moral restraint," that he had given short shrift in 1798. This introduced a somewhat hopeful note into Malthus's vision of human prospects, which remained nevertheless profoundly pessimistic.[35]

Although much criticized, Malthus's views quickly entered the mainstream of British thought and exercised a "magisterial influence on public opinion" until the last decades of the nineteenth century.[36] They became, as Robert Young has put it, "as commonplace in the first half of the nineteenth century as Freud's were in the twentieth."[37] Malthus enjoyed several advantages in the battle against his critics: his argument was easily comprehensible and lucidly presented, it provided the British ruling class with an argument against social reform, and it appeared prophetical after an 1801 census suggested that Britain had undergone a population explosion.[38]

Malthus's *Essay*, then, was an eminently reputable source when, in September 1838, while pondering possible mechanisms of evolution, the twenty-nine-year-old Charles Darwin picked up its sixth edition. He was much impressed and, as Edward Manier has demonstrated, elevated Malthus to an important place in the cognitive "cultural circle" that provided him with ideas about mind, social philosophy, religion, and scientific methodology. Malthus was the sole political economist among the forty-five authors cited most frequently by Darwin in his notebooks and early manuscripts; he was one of only six authors to be cited ten or more times there as well as in Darwin's autobiography.[39]

It would be unnecessarily tedious to review Darwin's many references to Malthus in subsequent years. These began with a note on September 28, 1838, and included substantial passages in Darwin's sketches of the selection theory in 1842 and 1844. In the 1850s, while laboring over his "big book" on natural selection, Darwin noted that he had also read Godwin's treatise on population and "various other Essays written against Malthus' great work, with all the attention of which I am capable, but I cannot say that they have had any weight with me."[40] Although he had earlier encountered the term "struggle for existence" in the work of geologist Charles Lyell, and although he knew that Benjamin Franklin and others had discussed the rapid rate at which organisms multiplied, Darwin always associated his struggle for existence with "the law of increase so philosophically enunciated by Malthus."[41]

When his *Origin* came under attack, Darwin identified with the unfairly maligned Malthus. Referring to one critic of the selection theory, he confided to Charles Lyell in June 1860 that "It consoles me that [he] sneers at Malthus, for that clearly shows, mathematician though he may be, he cannot understand common reasoning. By the way what a discouraging example Malthus is, to show during what long years the plainest case may be misrepresented and misunderstood."[42] When Alfred Russel Wallace urged him to abandon the confusing term "natural selection," Darwin responded, "I doubt whether the use of any term should have made the subject intelligible to some minds, clear as it is to others; for do we not see even to the present day Malthus on Population absurdly misunderstood? This reflection about Malthus has often comforted me when I have been vexed at the misstatement of my views."[43]

Darwin commented publicly on his debt to Malthus in a famous passage in his autobiography (1876):

> In October 1838, that is, fifteen months after I had begun my systematic enquiry, I happened to read for amusement Malthus on *Population*, and being well prepared to appreciate the struggle for existence which everywhere goes on, from long-continued observation of the habits of animals and plants, it at once struck me that under these circumstances favourable variations would tend to be preserved, and unfavourable ones to be destroyed. The result of this would be the formation of new species. Here, then, I had at last got a theory by which to work ....[44]

Other leading British evolutionists found Malthus an equally authoritative and reputable source. Alfred Russel Wallace indicated several times that Malthus's insights had played an important part in his own discovery of the selection theory. In the first of these reminiscences he recalled that he was living in Ternate and suffering from intermittent fever:

> During one of these fits, while again considering the problem of the origin of species, something led me to think of Malthus' Essay on Population (which I had read about ten years before), and the "positive checks,"—war, disease, famine, accidents, etc.—which he adduced as keeping all savage populations nearly stationary. It then occurred to me that these checks must also act upon animals, and keep down their numbers.... While vaguely thinking how this would affect any species, there suddenly flashed upon me the idea of the survival of the fittest—that the individuals removed by these checks must be on the whole, inferior to those that survived. Then, considering the variations continually occurring in every fresh generation of animals or plants, and the changes of climate, of food, of enemies always in progress, the whole method of specific modification became clear to me, and in the two hours of my fit I had thought out the main points of the theory.[45]

T. H. Huxley, renowned as "Darwin's Bulldog," was equally unapologetic about the Darwin-Malthus connection. Explaining the struggle for existence to a working-class audience, he commented:

> It is indeed simply the law of Malthus exemplified. Mr. Malthus was a clergyman, who worked out this subject most minutely and truthfully some years ago; he showed quite clearly,—and although he was much abused for his conclusions at the time, they have never yet been disproved and never will be—he showed that in consequence of the increase in the number of organic beings in a geometrical ratio, while the means of existence cannot be made to increase in the same ratio, that there must come a time when the number of organic beings will be in excess of the power of production of nutriment, and that thus some check must arise to the further increase of these organic beings.

And further:

> I know nothing that more appropriately expresses this, than the phrase, "the struggle for existence;" because it brings before your minds, in a vivid sort of way, some of the simplest possible circumstances connected with it. When a struggle is intense, there must be some who are sure to be trodden down, crushed, and overpowered by others; and there will be some who just manage to get through only by

the help of the slightest accident. I recollect reading an account of the famous retreat of the French troops, under Napoleon, from Moscow. Worn out, tired, and dejected, they at length came to a great river over which there was but one bridge for the passage of the vast army. Disorganized and demoralized as it was, the struggle must certainly have been a terrible one—everyone heeding only himself, and crushing through and treading down his fellows.[46]

Perhaps intrigued by Darwin's references to Malthus's *Essay*, the British botanist Sir Joseph Hooker read the work sometime in the 1860s. He shared his reaction with a North American colleague:

Did you ever read that painful book, Malthus on Population? I did the other day, and was painfully impressed by it. I had supposed he was a sort of materialist, who *advised* the checking of the population by restrictive means, and was surprised to find nothing of the sort, and a rather fine exordium at the end on a future state and the benefits of Christianity! His arguments seem incontrovertible to me.[47]

It has become a truism—and the above citations bear it out—that British evolutionary thought was related to broader British discussions of political and philosophical issues concerning man's place in nature (and what to do about it). It is one thing, of course, to identify this relationship and quite another to determine its precise impact on the thinking of any individual. This latter issue is so complex, and its resolution so closely related to one's assumptions about the nature of scientific thought, that the careful and energetic work of Darwin scholars—however much aided by a rich stock of manuscript material—has generated a variety of interpretations.

The issue for us here is much simpler. For our purposes it is not the creative but the associational dimension of the Darwin-Malthus connection that is of interest. In other words, what aspects of Darwin's theory might have been plausibly associated with Malthusian ideas by a contemporary reader, especially one interested in separating Darwin's scientific contribution from Malthus's political economy? There is a certain parallelism between this imaginary reader's attempt to distinguish between good science and bad cultural baggage, on the one hand, and the historian's effort to determine what, precisely, Darwin learned from Malthus, on the other. With this limited goal in mind, let us briefly review the conclusions of historians with an eye, not to the differences among them, but to the sum of their observations.[48]

The following citations do not do justice to their authors' conclusions, but they do illustrate the frequent identification of a general relationship between Darwin's theory and broader, specifically British developments:

"The theory of natural selection and the struggle for existence of Darwinian evolution . . . was a reflection of the free competition of the full capitalist era." (J. D. Bernal, 1954)[49]

Malthus's "theory and its assumptions about nature were at once pervasive in the [British] biological literature of the first decades of the century and a part of an ongoing debate within natural theology which was at least as important to Darwin and Wallace as the question of the mechanism of evolution." (Robert Young, 1969)[50]

The development of Darwin's thinking was "a microcosm of the more general development from a philosophy of nature and man appropriate for an agrarian and aristocratic world to one suitable for the age of industrial capitalism." (Dov Ospovat, 1981)[51]

"Darwin's evolutionary biology reflects a characteristically British intellectual outlook in its conception.... In the *Origin of Species* biology and economics 'joined hands' or perhaps more accurately biology joined hands with Scottish political economy, sociology, and historiography, and with English philosophy of science. The political economy was that of Adam Smith and his disciples." (Sylvan Schweber, 1985)[52]

As for Malthus's precise impact on Darwin's thinking, some historians, such as Ernst Mayr, acknowledge only a "very limited" and highly specific influence. For Mayr, Darwin borrowed Malthus's "'populational arithmetic,' but not his political economy," and did so only when his own examination of the writings of animal breeders had already led to the most significant shift in his thinking. He compares Malthus's role to that of "a crystal tossed into a saturated fluid."[53] Howard Gruber reaches a similar conclusion. For Gruber, Darwin found in Malthus a clear, mathematical statement of the "superfecundity principle," and did so at "just the right moment: when Darwin's thought had at last grown to the point where, having met natural selection so often, he could finally recognize it."[54]

Other scholars identify other dimensions of Malthus's influence. Sandra Herbert observes that, while Darwin had previously read about the interspecific struggle for existence—for example, in Lyell's *Principles of Geology*—Malthus impressed on him the importance of intraspecific conflict.[55] David Kohn indicates three important consequences of Darwin's reading of Malthus: First, Malthus "radically changed Darwin's attitude toward the balance of nature," shifting his attention from "the final equilibrium prescribed by the balance of nature to the dynamic process that creates the equilibrium." This led to Darwin's sudden and novel insight into the link between extinction and origin in the struggle for existence, an insight that "was not the predictable outcome of a gradual development in Darwin's thought."[56] Second, Kohn agrees with Herbert that Malthus exposed Darwin for the first time to well-documented cases of intraspecific competition. Third, "for Darwin, after Malthus, adaptation became a matter of sharp-edged wedges pounded into narrow cracks. Adaptation was no longer a temperate, quiescent process. It became the result of a fierce contest that differentiated among the well adapted, where even 'a grain of sand turns the balance.'"[57] An essay coauthored by Kohn and M. J. S. Hodge elaborates this point: "Shown by Malthus how to construe intraspecific competition as analogous to interspecific, Darwin could take very small differences in structural characters to determine the outcome of both."[58]

Much of this discussion was stimulated by Robert Young's "Malthus and the Evolutionists: The Common Context of Biological and Social Theory" (1969). Exploring the varied ways in which leading British evolutionists appropriated and interpreted Malthus's ideas, Young argued that Darwin transformed Malthus's doctrine into an evolutionary theory by emphasizing population pressures and removing Malthus's idea of moral restraint (which was irrelevant, of course, to plants and animals). Darwin "was, in effect, reverting to the purity of the inescapable dilemma of Malthus's first edition."[59] In later articles Young agrees that the actual role of Malthus's theory in Darwin's creative process was "very complex and open to a number of interpretations" but observes that scholarship about this role has confirmed his original thesis that British evolutionary thought was embedded in British discussions of man's place in nature.[60]

In the light of present-day scholarship, then, it would not be surprising if Darwin's contemporaries, especially those outside of the British cultural context, associated his struggle for existence with specifically British, bourgeois, or Malthusian values. This is especially true of three specific elements of Darwin's concept: (1) the term "struggle for existence" and much of the rhetoric associated with it, (2) the reliance on population pressures and a specific populational arithmetic to fuel this perpetual struggle, and (3) the emphasis on intraspecific conflict.

Selecting among Darwin's Metaphors

Darwin scholars have correctly insisted that Darwin's creative process and vision cannot be fully appreciated by isolating one of his "figures of thought" but only by examining their interaction and summary impression. The Malthusian tones of the "struggle for existence," for example, were modified by the benign interdependence of the "entangled bank," and these were joined by other important images such as "natural selection," "contrivances," "wedges," and the "great tree of nature."[61]

In his review of Darwin's system of metaphors, Howard Gruber notes the "remarkable differential uptake of Darwin's ideas." He observes that "later writers—including social Darwinists—have put almost the entire emphasis on the images of artificial selection and war." Presumably referring to English-speaking intellectuals, he adds:

> This biased account of Darwin's thinking has had two unfortunate consequences. First, it has given to the idea of struggle the form of *polarized struggle* between two opposed forces (as in human warfare, or as in a contest between Breeder and Nature). In Darwin's own thinking, "struggle" clearly means something else: the complex interplay of many factors, leading to the differential survival of organisms, depending on their varying adaptation to all the conditions of life.
>
> Second, this biased account of Darwin's thinking puts all the emphasis on forces of selection and destruction in nature, those aspects of the entire process in which stronger organisms flourish and weaker ones die, depicting a world of nothing but winners and losers. But this selective use of Darwin's ideas destroys the dialectical unity of his thought. The creative and explosively productive metaphors of contrivance, tangled bank, and branching tree were equally essential features of Darwin's image of nature.
>
> It would be worth studying the social and ideological sources of this distortion. The differential uptake of Darwin's ideas reflects a rather sour view of nature, as compared with the exuberance of Darwin's feeling and thought about it.[62]

Darwin himself, of course, gave his readers ample reason to assign greater weight to the struggle metaphor than to that, say, of the entangled bank. The former, after all, graced the subtitle of the *Origin* and the title of its third chapter. Furthermore—and notwithstanding Darwin's actual interpretation of the "struggle for existence"—his explicit association of that metaphor with Malthus could only have encouraged a particular reading of that phrase. Yet the point remains that the "differential uptake" and interpretation of Darwin's metaphors reflected a particular intellectual bent—Gruber calls it "sour"—among influential British thinkers.

A related point emerges from Edward Manier's analysis of the Darwin-Malthus connection. Like Gruber he laments the common misunderstanding of Darwin's image of nature, attributing this to "the failure of Darwin's readers to note (and Darwin's failure to insist upon) the distinction of his use of 'struggle for existence' and Malthus's use of any comparable term."[63] Manier demonstrates that Darwin did not simply appropriate Malthus's metaphor, but rather transformed it. For Malthus "struggle" was a "zero-sum competition for a scarce resource (subject to the law of diminishing returns)." For Darwin it became "an effort to overcome a difficulty (through relations of dependence, chance, variation or competition)." Malthus's essay, then, provided "a causally stimulating occasion for Darwin's inventive elaboration of the concept":

> The links connecting Malthus and Darwin were both causal and symbolic.... It is not the case ... that Darwin either affirmed or denied what he read in Malthus. Instead, I claim, Darwin critically reinterpreted and expanded or developed what had been a simple univocal term, "struggle," and constructed a complex (multivocal) metaphor of his own, in which three distinct senses of ["struggle"] were placed in interaction with each other. Communication of this degree of subtlety and precision requires a significant degree of disciplinary or cross-disciplinary agreement, but it does not require complete consensus.[64]

What bears emphasis here is that Darwin's transformation of Malthus's term "struggle" followed from an essentially friendly reaction to Malthus's theory. That is, Darwin was sufficiently sympathetic to Malthus's general outlook to identify substantial merit in his insights and to elaborate them for his own purposes. I think it a safe assumption that this creative process would have been quite different had Darwin found Malthus's ideas totally foreign to his own experience and ideology or had he been a resolute anti-Malthusian. Had the same been true of other leading British evolutionists, their reaction to Darwin's "struggle for existence" probably would have also been quite different.

This brings us to the question of the cross-cultural transmission of metaphors in scientific thought. If a metaphor is culturally dependent, as Darwin's "struggle for existence" clearly was, what happens to it under the scrutiny of thinkers who do not share the cultural assumptions that lend it commonsensical power and explanatory appeal?

The Russian Audience

The Russian empire of Darwin's day sprawled over about one twenty-fifth of the earth's surface and one-sixth of its dry land. Siberia alone occupied forty times the area of Great Britain and Ireland. This great mass was dominated by a seemingly endless plain that, interrupted only ineffectually by the Ural Mountains, extended from central Europe well into central Asia, and from the Far North to the Black Sea. Great mountain ranges rose in the empire's outermost regions—the Carpathians to the southwest, the Caucasus farther east between the Black and Caspian seas, and the Pamir, Tian Shan, and Altai ranges still farther east in Siberia.

Russia's severe continental climate has been described succinctly by one historian:

> Northern and even central Russia are on the latitude of Alaska, while the position of southern Russia corresponds more to the position of Canada in the western hemisphere than to that of the United States.... In the absence of interfering mountain ranges, icy winds from the Arctic Ocean sweep across European Russia to the Black Sea. Siberian weather, except in the extreme southeastern corner, is more brutal still. In short, although sections of the Crimean littoral can be described as the Russian Riviera, and although subtropical conditions do prevail in parts of the Southern Caucasus, the overwhelming bulk of Russian territory remains subject to a very severe climate. In northern European Russia the soil stays frozen eight months out of twelve. Even the Ukraine is covered by snow three months every year, while the rivers freeze all the way to the Black Sea. Siberia in general and northeastern Siberia in particular belong among the coldest areas in the world.... Still, in keeping with the continental nature of the climate, when summer finally comes—and it often comes rather suddenly—temperatures soar. Heat waves are common in European Russia and in much of Siberia, not to mention the deserts of Central Asia which spew sand many miles to the west.[65]

This inclement continental expanse was but thinly populated. In 1875 the Russian geographer P. Belov estimated the empire's total population at about 82 million: 65.5 million in European Russia, 10 million in Asian Russia (Siberia, Turkestan, and Transcaucasia), 5 million in Poland, and 2 million in Finland. Especially compared with western Europe, Russia was a land of open spaces. This circumstance, Belov observed, had created some problems—for example, the national economy suffered from the difficulty of communicating across such great distances—but it had also spared Russia "those evils which afflict states with little land and a surplus population that often has difficulty obtaining the means of subsistence."[66]

The great majority of Russians indeed had difficulty obtaining food, clothing, and shelter—the empire was notorious for the poverty and poor health of its subjects—yet neither Belov nor other observers could plausibly attribute this state of affairs to population pressures on scarce resources. Even the phenomenon of peasant "land hunger," which gained attention amid a rapid population increase during the last half of the nineteenth century and which would seem, at first glance, to lend itself to a Malthusian interpretation, was analyzed in quite different terms by Russians who pondered it against the backdrop of their homeland's great expanses.[67]

The publication of Darwin's theory coincided with a tumultuous period in Russian social and intellectual history. Between the appearance of the first English edition of the *Origin* in 1859 and the first Russian edition in 1864, Tsar Alexander II implemented a series of fundamental reforms. Most important was the emancipation of the serfs in 1861, which affected some fifty-two million peasants, sounded the death knell of feudal relations, and accelerated the development of capitalism.

Russia's two great social classes remained the landlords and the peasantry, whose relationship was mediated by the peasant commune, or mir. Through the mir, peasants collectively met their obligations to the landlord (or, on state lands, the government) and periodically redistributed and equalized land among its members. This was unchanged by the emancipation decree of 1861, which did not grant peasants the right of private land ownership and required them to belong to a commune, through which they made their redemption payments.

Russia's bourgeoisie was weak, timid, and highly fragmented. As for the proletariat, in the 1860s and 1870s Russian intellectuals could still debate whether it really existed in their country. The consolidation and expansion of an urban working class awaited Russia's industrial revolution, which arrived in the late 1880s and 1890s, about one century after Great Britain's.[68]

This class structure was conducive to the development of neither a significant pro-laissez-faire political tendency nor, until the turn of the century, a broad-based Marxist movement. Political discourse was dominated by two general outlooks: monarchism and a socialist-tending populism. Representatives of each exhibited a self-conscious ambivalence toward western Europe, particularly Great Britain. They respected its vitality, its science and technology, and its economic strength. These were often contrasted to Russian inertia, backwardness, and illiteracy, and never more bitterly than after Russia's humiliating defeat in the Crimean War (1854-56). Yet western Europe, and Great Britain especially, was also perceived as unacceptably philistine, individualistic, and socially incoherent. Russia's "Westernizers" and "Slavophiles" expressed a panorama of views concerning the possibility and means of importing Western virtues while embargoing Western vices.

One could find a mirror image of these sentiments in the writings of such widely read western European observers of Russian life as August von Haxthausen (1850) and Georg Brandes (1889). They portrayed an economically and technologically primitive land, but one with communitarian virtues that western Europeans might well envy. Writing in the wake of the defeated European revolutions of 1848, Haxthausen described a cohesive society that had preserved a unity of classes unknown in western Europe. As S. Frederick Starr has observed, Haxthausen comforted defeated Continental radicals with his message that "the communitarian ideal which had just been discredited on the streets of Paris, Berlin, and Vienna was alive and well on the eastern border of Europe."[69] Continental conservatives, too, were heartened by Russia's alleged achievement of social harmony within traditional social institutions. Haxthausen's volume found its toughest audience in Darwin's homeland, in "liberal, industrial England," where his homage to Russian communitarianism encountered "skepticism, criticism, and outright derision."[70]

The social transformation of tsarist Russia in the 1860s contributed to an explosion of interest in natural science and a qualitative growth in scientific institutions. Radical theorists such as D. I. Pisarev and N. G. Chernyshevskii hailed positive knowledge as a powerful weapon in the battle against the metaphysical obfuscations of tsarist ideology and a key force for social progress. Youthful radicals, many of them the sons and daughters of declining gentry families, found science an ideologically satisfying vocation appropriate to post-reform society. Embryologist A. O. Kovalevskii and his brother, paleontologist V. O. Kovalevskii, zoologist and pathologist I. I. Mechnikov, physiologists I. P. Pavlov and I. M. Sechenov, and plant physiologist K. A. Timiriazev were among the scientific community's important recruits in these years. The tsarist state lent critical support ("critical" in both senses) to the growth of scientific institutions. Suspicious of materialist currents in Western science, particularly in physiology and evolutionary biology, state authorities nevertheless thought a native scientific cadre essential to their economic, medical, and military goals. So they often subsidized the scientific training of "subversive" students and scholars in western Europe, published their ideologically "pernicious" works, paid their salaries, and financed an expanding system of university science departments, laboratories, and scientific societies.[71]

Darwin's theory was first communicated to the Russian public in January 1860, when the *Journal of the Ministry of National Education* published a translation of Lyell's favorable comments about it before the British Association for the Advancement of Science. His ideas reached the broader reading public in 1864 with the publication of S. A. Rachinskii's translation of the *Origin* and the appearance of Pisarev's and Timiriazev's popular essays on the selection theory. Rachinskii's edition sold out quickly, and others soon followed. Darwin's other works were also translated rapidly and read eagerly: the Russian and English editions of *Descent of Man* (1871) appeared almost simultaneously, and the Russian publisher of *The Expression of the Emotions in Man and Animals* (1872) actually beat the English to press (V. O. Kovalevskii accomplished this feat by using plates from his previous edition of a work by Brehm). A translation of *The Variation of Animals and Plants under Domestication* (1868) appeared in the early 1870s.

The reaction of Russian intellectuals to Darwin's theory was, of course, uneven, but generally quite favorable. For the great majority Darwin became a highly prestigious figure—the embodiment of modern natural science, the author of a powerful argument for evolutionism, and the discoverer of an important factor in evolution, natural selection. As A. O. Kovalevskii recalled in 1909:

> Darwin's theory was received in Russia with profound sympathy. While in Western Europe it met firmly established old traditions which it had first to overcome, in Russia its appearance coincided with the awakening of our society after the Crimean War and here it immediately received the status of full citizenship and ever since has enjoyed widespread popularity.[72]

In 1896 M. A. Antonovich observed similarly that Russian scholars had reacted with "complete sympathy" to Darwin's theory, welcoming it as "long-awaited guest." He added, however, that after initial silence or mild approval, conservative "ignoramuses and self-styled patriots" launched an attack upon the selection theory in the late 1870s.[73]

This generally favorable reception has long been acknowledged by historians, as have several reasons for it: Russia lacked an entrenched creationist tradition and had already produced a number of evolutionists (most notably zoologists K. F. Rul'e and K. E. von Baer, but also A. N. Beketov, A. P. Bogdanov, K. F. Kessler, N. A. Severtsov, G. E. Shchurovskii, and others). Furthermore, evolutionism resonated with the materialist and positivist inclinations of the urban intelligentsia of the 1860s, particularly of the progressives who flooded the biological sciences.[74]

Darwin's Russian audience was, then, well prepared for his evolutionism. It was, by the same token, well prepared to evaluate critically the novel features of Darwin's selection theory, particularly his appeal, through the metaphor "struggle for existence," to a body of common knowledge that proved not so common at all.

CHAPTER 2

Malthus, Darwin, and Russian Social Thought

The Malthusian theory ... has always been indignantly rejected by Russian economists.[1] (K. A. Timiriazev)

The wretched pastor Malthus and the great naturalist Darwin! What an original and unexpected combination of names![2] (P. N. Tkachev)

In 1799, while gathering information for the much-revised second edition of his *Essay on the Principle of Population*, the Reverend Thomas Robert Malthus visited Russia. He was impressed by the "extraordinary fertility" of Southern Siberia and by its "deficiency of population" relative to the land's agricultural potential.[3] "Notwithstanding the facility with which ... the most plentiful subsistence might be procured," he observed, "many of these districts are thinly peopled." This demonstrated that "the nature of government, or the habits of people" could present such obstacles to the production of food that "a part of the society may suffer want, even in the midst of apparent plenty."[4]

Malthus noted approvingly that Russian tsars had attempted to increase the country's population by introducing manufacturing and aiding colonists in distant regions, and surmised that these efforts had been mildly successful.[5] Yet the fundamental check on population growth was Russia's feudal system itself:

> Russia has great natural resources. Its produce is, in its present state, above its consumption; and it wants nothing but greater freedom of industrious exertion, and an adequate vent for its commodities in the interior parts of the country, to occasion an increase of population astonishingly rapid. The principal obstacle to this, is the vassalage, or rather slavery, of the peasants, and the ignorance and indolence which almost necessarily accompany such a state.[6]

Malthus, then, saw an expansive, thinly populated land that, once freed of feudal social relations, could provide comfortably for a vastly increased number of inhabitants.

This reaction dramatizes the distance between Malthus's basic insight and Russians' experience. For a Russian to see an inexorably increasing population inevitably straining potential supplies of food and space required quite a leap of imagination.

So distant from Russian reality, Malthus's *Essay* was not even reviewed in a Russian journal until 1818, twenty years after its publication. Not until 1868, by which time Darwin had drawn attention to it—and sixty years after its translation into French, German, and Spanish—did the *Essay* appear in a Russian edition.[7] It has appeared in that language only once more to this day, in a much-abridged edition of 1895.[8]

As a description of Russian reality, then, Malthus's *Essay* offered little of interest.[9] As a political document, however, beginning in the 1840s, it was sharply criticized. Western Europe was a yardstick by which Russian intellectuals measured their own aspirations, and they discussed Malthus while evaluating British life and thought. Radicals agreed with monarchists that Malthus's progressions were but an arithmetical illusion reflective of an inhumane and soulless individualism. Radicals, who hoped to build a socialist society, saw Malthusianism as a reactionary current in bourgeois political economy. Monarchists, who hoped to preserve the communal virtues of tsarist Russia, saw it as an expression of the "British national type." Religious thinkers in the Orthodox church often added that Malthus was a heretic who blasphemously accused God of implanting a contradiction within the natural order.

By tying his concept of the struggle for existence to Malthus, then, Darwin may have gained plausibility with his British audience, but he almost assured the skepticism of his Russian readers.[10]

The Russian Reception of Malthus, 1798–1864

Malthus's *Essay* was first reviewed for a Russian audience in the conservative journal *Syn otechestva* (*Son of the Fatherland*) in 1818. An anonymous columnist concluded his relatively sympathetic appraisal with the observation that Malthus's principles were "very useful in the current situation in England," but were, of course, inapplicable to Russia.[11] The following year K. F. German published his monumental statistical survey of the Russian empire. In it he observed that "Russia can feed 960 million inhabitants, that is, almost the entire population of the world today." His discussion of population dynamics ignored Malthus entirely.[12]

The next important reference to Malthus appeared in 1844 in a demographic work by K. I. Arsen'ev, a liberal professor at the Central Pedagogical Institute in St. Petersburg who had studied population dynamics since the 1810s. Expressing the common view that population growth was essential to the wealth of a nation, Arsen'ev's 1844 study portrayed Russian history as a long battle against underpopulation and the poverty it engendered. In the fifteenth century, he observed, a traveler from Astrakhan to Moscow would pass through only two cities, traverse thousands of miles of virtually uninhabited forest, and find only thirty thousand inhabitants in the capital. Until the eighteenth century "Russia presented a picture of sparse population, poverty and weakness."[13] Peter the Great's modernizing policies had begun to improve the situation. From 1720 to 1835 the nation's population had almost quadrupled; now "Russia's greatness, its power, the rapid development of its strength and its gigantic successes amaze friend and foe alike."[14]

Malthus enjoyed great fame in England, Arsen'ev explained, and had been the first to argue that "population size is necessarily in harmony with the quantity of food supplies, or that there cannot be more inhabitants in a nation than there are supplies to feed them."[15] But this contention ignored the fact that additional people produce the additional "means of existence" that they require. He added:

> Malthus's proposition, one-sided and very limiting for other powers, would be beneficial for Russia, since, judging by its natural riches and the breadth of its cultivable lands ... it could sustain a population far surpassing that of the remainder of Europe. A seductive idea for the Russian State, but is it practicable?[16]

Malthus's progressions clearly held no terror for Arsen'ev.

Malthus did have one vociferous Russian defender, the economist A. I. Butovskii. In his *Essay on National Wealth, or On the Principles of Political Economy* (1847), Butovskii praised Malthus as the "Galileo of political economy" who "had first thrown light on the great laws to which population is subject."[17] Malthus had demonstrated "the constant tendency of peoples, carried away by natural instincts, to multiply beyond the means which, in the given conditions, the most enlightened and inventive labor, with the aid of the most enormous capital, can extract from nature for their subsistence." He had proven that poverty, premature death, low wages, high prices, and crime resulted from the imbalance of population and food supply, and had demonstrated that human salvation lay in moral restraint, especially on the part of laborers.[18] For these positions Malthus had been roundly vilified, but "his sin consists only in this—that he first discovered, and first resolved openly to shed light upon, the fatal aspect of a system that recognized the right of the poor to state assistance."[19]

Yet even Butovskii never analyzed Russia's problems from a Malthusian perspective. Addressing the future of the peasant commune in an article of 1858, he conceded that Malthus's principles were, of course, inapplicable to "our broad and expansive Russia."[20]

The most thorough and influential critique of Butovskii's *Essay on National Wealth* was socialist V. A. Miliutin's essay "Malthus and His Opponents" (1847). For Miliutin, Malthus's belief that humanity faced "eternal suffering and torment" rested on two fundamental errors.[21] First, by isolating population dynamics from broader social circumstances, Malthus had mistaken a phenomenon peculiar to "imbalanced" societies for a universal truth.[22] His ready generalization from English conditions, and the popularity of his doctrine, reflected its class bias: "Malthus was a zealous defender of Toryism, an economist of the privileged classes at a time when the aristocracy had exhausted, without success, all the resources at its disposal for resisting the pressure of new ideas and popular demands."[23] A rationalization for poverty and inequality in capitalist England, Malthus's doctrine would not apply in a harmonious, socialist society. Socialism would cultivate a balance between population and production, and would exploit the essentially limitless productive potential of the land. Population growth would stimulate technological innovation and a further division of labor; the resulting increase in production would more than compensate for any additional mouths to feed.[24]

A second weakness in Malthus's theory was its violation of the "belief in the immutable rationality of nature, its constant, and infallible ... subordination of everything to general, unified laws which establish harmony and order in all phenomena of the physical and moral world."[25] This teleological principle provided "the point of departure for every scientific investigation."[26] Malthus's advocacy of reproductive restraint violated the biblical injunction to "be fruitful and multiply," and reflected the heretical view that "there were established in nature, at one and the same time, two laws that were directly opposed to one another and by their very essence were unavoidably in constant struggle."[27]

The left held no monopoly on anti-Malthusianism. The monarchist Prince V. F. Odoevskii excoriated Malthus in his popular novel *Russian Nights* (1844). For Odoevskii, Malthusianism expressed the cardinal vices of British political thought—its narrow utilitarianism, its lack of a cohesive social ethic, and its permeation by "the coarse materialism of Adam Smith."[28]

> The country which wallowed in the moral bookkeeping of the past century was destined to create a man who focused in himself all the crimes, all the fallacies of his epoch, and squeezed strict and mathematically formulated laws of society out of them. This man, whose name ought to be preserved for posterity, made a very important discovery: he realized that nature had made a mistake by endowing mankind with the ability to multiply, and by not knowing how to make people's lives conform to their housing. This profound man decided that nature's mistake ought to be corrected, and its laws sacrificed to the phantom of society.
>
> "Rulers!" he used to exclaim in his philosophic rapture, "my words aren't just an empty theory; my system is not just a result of intellectual speculations. It is based on two axioms. The first: man has to eat, the second: people multiply. You do not contradict? ... Look what I have discovered: if your state were to prosper, if it were to enjoy peace and happiness, its population will have doubled in twenty-five years; in the next twenty-five years it will have doubled again, then again and again. Where will you find means in nature to provide food for them? ...
>
> "Don't delude yourselves with dreams about the wisdom of Providence, about virtue, love toward mankind, and charity. Try to understand my calculations: whoever comes late into this world has no place at the feast of nature; his life is a crime. Hasten to prevent marriages; let depravity destroy whole generations in their embryonic state; care not about the happiness of your people and of peace. Let wars, pestilence, the cold and mutinies destroy the faulty decrees of nature. Only then can the two progressions fuse, and the crimes and misery of each member of society will make possible the existence of a society itself."[29]

Odoevskii portrayed an economist who, possessed by the new scientific spirit, "worshipped Malthus and constantly filled sheets of paper with statistical calculations."[30] Driven to delirium by "the illness of his discontented reason," he left a manuscript, entitled "The Last Suicide," which described the nightmarish future facing mankind.[31] Odoevskii's Faust was not appeased by Malthus's later revisions of his *Essay*. These were mere contrivances designed to seduce well-meaning readers by concealing the essential Malthusian logic. There was but one consolation: "Malthus is the last absurdity in mankind; one cannot go any further in that direction."[32]

Russian critiques of Malthusianism became more numerous in the period known as the "sixties," which began with Alexander II's ascension to the throne in 1855 and ended about one decade later. These years were characterized by broad social ferment and sharp political debate, and by the implementation of several important social reforms, most notably the emancipation of the serfs in 1861. Malthus's principle of population was drawn into discussions of the relative merits of feudalism, capitalism, and socialism—and was rejected by the Left and most of the profeudal Right. Some conservatives argued that Malthus's law was applicable to capitalist societies and that Russia's feudal institutions should be preserved as an obstacle to it; some liberals argued that overpopulation might eventually prove a problem in the absence of industrial growth. Even these few thinkers, however, denied the inexorability of Malthus's progressions and disassociated themselves from Malthusian fatalism and social prescriptions.[33] So, just prior to the publication of the Russian edition of *Origin of Species* (1864), Malthus was sharply attacked and only weakly defended. Furthermore, the nature of this weak defense was to disavow Malthus's view that population pressures inevitably led to conflict, poverty, and suffering.[34]

The most influential radical journal of the first half of the decade was *Sovremennik* (*The Contemporary*). Its leading theorist, N. G. Chernyshevskii, criticized Malthus in several essays, most extensively in his commentary on John Stuart Mill's *Principles of Political Economy*.[35] Although the censor suppressed a chapter of this essay entitled "An Interpretation of the Meaning of the Malthusian Theory," the published comments consolidated and sharpened the standard Left critique of Malthusianism. Chernyshevskii identified fundamental mathematical errors in Malthus's analysis and attributed them to Malthus's desire to blame nature for the depredations of class society. "When purchasing power is in the hands of one man and hunger in the stomach of another," Chernyshevskii insisted, "then food for the latter will not be produced, although nature presents no obstacles for its production."[36] Radicals, liberals, and conservatives alike—including both Russian translators of the *Essay On Population*, P. A. Bibikov and I. A. Verner—praised Chernyshevskii for his definitive refutation of Malthus's theory.

After Chernyshevskii's arrest and exile to Siberia in 1862, *The Contemporary* began to yield its dominant position within the Left to *Russkoe slovo* (*Russian Word*). Its chief theorist, D. I. Pisarev, lambasted Malthus in his "Essay on the History of Labor"(1862). Pisarev praised Chernyshevskii for exposing "the general untenability and individual blunders of the Malthusian theory" and agreed that Malthusianism was an attempt by elitist intellectuals to "legitimize and justify the unsightly phenomena of modern life."[37] This required them to distort the character of the natural world as well:

> The earth and its productive forces are likened by Malthus to a chest full of money; if, he reasons, the money is divided among five people each gets a fifth part; if it goes to ten, each gets only half as much as in the first case; if it goes to twenty, each gets a quarter, and so on. From this follows the conclusion that the fewer people lay claim to the money, the richer will be those who will divide the contents of the chest among themselves. . . . Malthus counts up the increase in the total amount of produce just as precisely as interest on definite monetary capital can be calculated. . . . All along there is his tendency to regard the productive forces of nature as a dead mass to be measured in feet and weighed in pounds. In human labour, also, he sees only the mechanical application of muscular force and completely disregards the activity of the brain, which is constantly triumphing over physical nature and discovering new properties in it.[38]

This "Malthusian attitude toward nature" was profoundly mistaken. All living forms were united by the "eternal rotation" of matter, in which the prosperity of one life form supported that of another. Plants provided food for herbivores, which in turn supplied the beasts of prey. If a population and its requirements increased, this exchange of resources sped up proportionally. For humans as for all organisms, then, increasing numbers brought not privation but an "abundance of forces."[39]

By the mid 1860s Malthus had been firmly established as a bête noir among radical thinkers. N. V. Sokolov, staff writer for *Russian Word*, wrote, "For both the Jesuits and the Malthusian school of economists the end justifies the means; the Jesuits lie and deceive in the name of the Catholic Church, and the economists do so in the name of capital, which for them has become a deity."[40] Sokolov's colleague V. A. Zaitsev added that "the ethics of the Malthusians are the ethics of swine."[41]

Russian reactions, then, revolved around a perception common to radical and conservative thinkers: Malthusianism reflected an atomistic and soulless ideology, rooted in British political economy and culture, that violated Russians' vision of a cohesive society in which all of its members were valued parts of the whole. A. I. Herzen summarized this broad consensus when he compared Malthus's values with those of the cherished peasant commune. The commune, he observed, embodied an economic principle that was "the perfect antithesis of Malthus's celebrated proposition: it allows everyone without exception to take his place at the table."[42]

The Radicals' Dilemma

The dilemma posed to radical thinkers by the Darwin-Malthus connection was put bluntly by P. N. Tkachev: What, he asked, could the "great Darwin" have in common with the "pastor thief"?[43] This connection—summarized so unavoidably in Darwin's statement that his struggle for existence was "the doctrine of Malthus applied with manifold force to the whole animal and vegetable kingdoms"—combined with other developments in natural science to expose serious weaknesses in radical thought.[44]

Leftist ideology in the 1860s rested heavily on a positivist faith in the natural sciences. As Pisarev expressed it, "Mankind has only one evil—ignorance; and against this evil there is only one medicine—science."[45] Natural science was free of ideological contamination and so was the only reliable source of social values. "If that social order in which we live is so bad for the majority," Zaitsev proclaimed in 1863, "this is because the natural sciences do not lie at its foundations, because human life is constructed [instead] . . . upon the basis of abstract ideas."[46] The biological sciences had a particularly important role to play. By demonstrating that "man is first of all an animal," they struck at the core of tsarist ideology, particularly at its tenet that an immaterial "spiritual aspect" separated man from the animal world and tied him to God, the Church, and the tsarist social order.[47]

Evolutionism was highly compatible with these views. N. V. Shelgunov was delighted to observe, in an essay on Lyell, Darwin, Vogt, and Huxley, that "contemporary natural science has once again included man in the very same category in which Linnaeus placed him one hundred years ago—man is in the same category of organisms as the bat."[48] Playing with the sensitivity of those offended by this conclusion, he added:

> We will not be disappointed that we, the builders of railroads, steamships, electrical telegraphs and other marvels of civilization, are in the same category as the flying lemur, because this does not at all obligate us to fly like a bat or make faces and live like monkeys; our human characteristics will remain with us and will only serve as evidence of the great results which can occur from the gradual accumulation of features advantageous for the development of a form, and of the improvement to the race which can result from such an accumulation.[49]

However comfortably evolutionism fit radical ideas, Darwin's concept of the struggle for existence, and his explicit debt to the reactionary Malthus, posed a serious problem. The scientism of the Left dictated that Darwin and Malthus be judged together. The laws of nature could not be fundamentally different from the laws of society. How, then, could scientific progress be so closely tied to an apologist for capitalist exploitation? And how could the Malthusian factors of overpopulation and intraspecific competition—these transparent rationalizations for exploitation and suffering—constitute the central agents of evolutionary change? This dilemma was exacerbated by the common belief that evolutionary theory dealt, not simply with adaptation, but with the laws of natural and social "progress."

There were two general responses to this problem. The dictates of natural science could be accepted and "sentimental values" reconsidered, or Darwin's theory could be criticized as partially or entirely false, as the distortion of science by bourgeois ideology.

Pisarev and Zaitsev chose the first alternative. Reviewing the *Origin* in an essay entitled "Progress in the Animal and Vegetable Worlds" (1864), Pisarev enthused: "In this theory the readers will find the rigorous precision of an exact science, the boundless breadth of philosophical generalization and, finally, the superior and irreplaceable beauty which is the mark of vigorous and healthy human thought."[50] Having dismissed the "Malthusian attitude toward nature" just two years before, he now urged his readers to consider the "amazing results" of the geometrical multiplication of organisms; were there not a constant struggle for existence, the earth would rapidly become overpopulated. Two years earlier Pisarev the anti-Malthusian had written lyrically of the "universal rotation of matter" and of mutual support in the natural world. Now Pisarev the Darwinist placed Malthusian overpopulation and struggle at the very center of nature. "The overwhelming majority of organic beings come into the world as into a huge kitchen where the cooks are chopping, disembowelling, stewing and frying one another."[51] He continued in his usual florid style:

> Organic life is inconceivable without the continuous destruction of living beings; organic life is an eternal struggle between living beings and every organic form is limited in its reproduction by all the other forms. This struggle cannot stop for an instant, for every step in life is an act of struggle. Everything has to be fought for: food, space, a handful of earth, a breath of air, a droplet of water.... Whoever makes a blunder in that struggle is doomed: he is scrapped like old-fashioned earrings or an old pin; he dies and is immediately devoured in the merriest and most

good-natured way by other plants and animals.... If, for instance, the hawk catches and kills a pigeon, it wins a victory not over the pigeon alone, but over other hawks too.... Hence there is a continual struggle among these birds even when it does not come to an open fight between them. If people are looking for mushrooms in the same wood there is obviously a struggle among them, although they exchange no blows.[52]

Zaitsev took this analysis one step further in a controversial article for *Russian Word* in 1864. Since the struggle for existence was the engine of progress in the natural world, it must also be socially beneficial. Pointing to biological differences between the races, Zaitsev reached a conclusion that was heretical for Russian radicals: "Slavery is the very best outcome for which a colored man could hope upon coming into contact with the white race," he explained. Appeals to the brotherhood of man were unscientific, worthy only of "a sentimental lady of the manor."[53] When M. A. Antonovich objected that "whatever zoology says, sane thought and the general welfare must be respected," Pisarev came halfheartedly to Zaitsev's defense.[54] Zaitsev later claimed that his position had been misinterpreted, but a serious problem with radical scientism had been irrevocably exposed.[55]

N. D. Nozhin chose the second alternative. He wrote with some authority on biological issues. Like many youths of the 1860s, he had been attracted to natural science by its prestige in the radical community. In the years 1861–1864, while in Heidelberg, Tübingen, and Messina, he had studied embryology and zoology. He became friendly with A. O. Kovalevskii and I. I. Mechnikov; the scientific research of each was influenced by Nozhin's attempt to discover "a general law for the arrangement of and relationship between the body tissues of all animals."[56] Returning to Russia in 1864, Nozhin presented his scientific work to the Academy of Sciences and addressed the relationship of science to social issues in several lectures and articles. He died suddenly in 1866, at the age of twenty-five.

Two months after Nozhin's death, his review of Darwin's theory appeared in *Knizhnyi vestnik* (*The Book Messenger*). Focusing on Darwin's Malthusianism, Nozhin argued that intraspecific relations were normally characterized, not by competition, but by mutual aid. For Nozhin evolutionism was an already firmly established scientific truth; the import of Darwin's theory lay elsewhere:

> We even think that the entire significance of Darwin's theory lies not in the question of the origin of species itself, but in the search for general biological laws of organic life, in *the search for laws of the metamorphosis of organisms*, and only this zoological question becomes a living issue, a social issue; ... an issue concerning the physiological and pathological laws of animal life and human societies.[57]

Darwin's approach to this key question identified him as a "brilliant bourgeois naturalist" whose theory "is true only in the sense that Malthus's theory is also true."[58] Bourgeois ideology had blinded Darwin, as it had Malthus, to the fact that the struggle for existence was not a normal, healthy, physiological process resulting in progress but merely an abnormal "source of pathological phenomena."[59]

Both Malthus's and Darwin's theories obtained only under "conditions of contact between different species or to relations between unintegrated individuals of one and the same species."[60] Nozhin drew upon his own expertise in zoology to emphasize this point:

> The hydra knows very well that each new consumer [bud] growing upon it is a worker and becomes, under the normal conditions of life, not only harmless but even very useful to it (which, of course, would not be the case if the buds had different needs, and were possessed of different tendencies). Each new member of the colony, that is, each newly formed bud, possessed of organs for grasping, and applying them to this task, earns much more then necessary for its own needs, and the inevitable surplus becomes the source of greater and greater means of life, and so leads to the formation of new buds. So, an increase in the number of buds does not generate any antagonism of interests between the component parts of the colony.[61]

Among well-integrated organisms, then, increasing numbers brought additional labor power and prosperity, not conflict and impoverishment. Malthus had mistaken the pathological exception for the norm. Similarly, Darwin had mistaken the "disturbed balance" of contemporary English society for a universal and beneficent natural law.[62] As was evident from Zaitsev's scandalous article on slavery and colonialism, Darwin's theory threatened to bring "new tears and sorrow to mankind."[63]

Most Russian radicals soon concluded that Pisarev and Zaitsev's approach to the Darwin-Malthus connection was bankrupt, and this embryonic "Social Darwinist" tendency was stillborn.[64] Nozhin's analysis of this troublesome relationship, on the other hand, foreshadowed the populist approach of subsequent decades.

The Bibikov Edition

Most Russians were unable to read Malthus's *Essay* until P. A. Bibikov translated the fifth edition in 1868. Widely reviewed in the popular press, this volume included a lengthy introductory essay that reflected and reinforced the perception of a fundamental connection between Malthus and Darwin. Bibikov's critique of Malthus rested on a close analysis of the struggle for existence, skepticism about the role of overpopulation as a source of struggle, and consideration of cooperation among like individuals—themes that became increasingly important in reactions to Darwin as well.

A contributor to *The Contemporary* and to Dostoyevski's journal *Vremia (Time)* in the 1860s, Bibikov had suffered a spectacular failure in his chosen career. In 1865 he had published his essays in a single volume whose aggressive positivism offended the censor. Bibikov was arrested and his volume confiscated. After some deliberation, however, the censors decided that the offending arguments were so poorly presented as to be harmless. The author and his volume were released.[65]

A more disheartening reaction to one's first book is difficult to imagine. Bibikov began a new career as translator and produced a series of thirteen volumes, including works by Francis Bacon, Xavier Bichat, and Adam Smith. He introduced each with a long essay weighing the strengths and weaknesses of the author's views.

Why bother to translate Malthus's *Essay*? It was well-known, Bibikov conceded, that this work was a reactionary attempt to refute radical theories about the perfectibility of man.[66] Despite its unsavory origins, however, the *Essay* contained an important kernel of truth. Bibikov assigned himself the task of identifying this relative truth and freeing it from the strictures imposed by Malthus's "reactionary purpose."[67]

He introduced Malthus's central proposition as follows: "All living entities, including plants, tend [*stremliat'sia*][68] to multiply without limit, and would actually do so if they did not confront inadequate space and food."[69] This basic principle seemed so self-evident that even Chernyshevskii, "the most talented, indisputedly best and sharpest of Malthus's opponents," had not attempted to refute it. Yet it was pure sophistry.[70]

To demonstrate this Bibikov turned to "the most powerful defender and sharpest investigator of this [Malthusian] natural law"—Charles Darwin. Although Darwin paid homage to Malthus, he had demonstrated that the multiplication of a species was actually limited by three different types of struggle: struggle within a species, struggle with other species, and struggle with the physical conditions of life.[71]

These three types of struggle often transpired in the same situation. One could not, therefore, assert that only intraspecific struggle prevented overpopulation. Yet it was precisely this assertion that lay at the heart of the Malthusian theory.[72] Bibikov illustrated this by interpreting a passage from the *Origin*:

> The mistletoe, a parasitical plant that lives on the apple tree, not only struggles for its existence with the climate and natural conditions, it struggles also with the apple trees, since the development of a great number of mistletoe on one tree will make the tree wither and die; but several mistletoe seedlings, sprouting one around the other on one tree branch, also struggle among themselves. Obviously, I have no right to say that the mistletoe struggle only among themselves for food, and that only this struggle stops their multiplication. I dare not forget that the mistletoe conduct still another struggle with organisms of other species . . . nor dare I forget the natural conditions that also prevent it from multiplying. But that is exactly what the Malthusian theory does.[73]

Malthus's error could be rectified by separating the struggle for existence into two categories: one "completely independent of competition" and the other "conditioned by competition."[74] When cats pursued mice, the conflict between these two species involved no competition. Yet the cats simultaneously competed with each other regarding their relative skills as predators, and the mice competed among themselves regarding which could best elude their foe.

What was the relationship between these different aspects of the struggle for existence? Bibikov contended that the weakening of one strengthened the other. "The more feeble the cats' attack, the more strenuous is the struggle for existence among the mice themselves; the more mice that succumb to the cats' attack the weaker is the war among them for the means of existence."[75] Intraspecific competition, then, was most intense when there was no pressing threat from other species and the physical conditions of life.

This logic exposed a fallacy in Malthus's central proposition. The British parson had assumed that population growth was limited only by the supply of food and space. But interspecific struggle—such as the cats' attack on the mice—was often the chief obstacle to population growth. A few years earlier, Bibikov recalled, millions of gophers had suddenly appeared in southern Russia. Their numbers had grown inexorably until, suddenly, they disappeared. The gophers had not exhausted their food supply, as was evident from the continued harvests of local cornfields. Rather, they had fallen victim to an interspecific struggle with intestinal worms.[76]

Malthus's preoccupation with the limitations imposed by food supply and intraspecific competition had led him to underestimate the importance of interspecific struggle, abiotic conditions, diseases, and other important checks on population. His central postulate, then, should be reformulated: "Organisms tend to multiply to the number permitted by the food supply, but can never attain that number due to other conditions that are disadvantageous for their multiplication."[77]

This new formulation could be fruitfully applied to human populations, which were subject to many checks. Chief among them were ignorance and poor social organization. Referring to Russia, Bibikov observed that population growth was surely not inhibited by "a natural, inevitable lack of nutriment" in a country that cultivated only one-third of its arable land, and only one-fifteenth with proper techniques.[78] Only when such checks to population growth were eliminated would Malthus's central proposition hold true. No organism had ever achieved such a happy state; for mankind to do so would present "the most propitious conditions for human happiness."[79]

For Bibikov this analysis of the struggle for existence resolved the tension between two truths of political economy: that struggle encouraged productivity and that it also led to poverty.[80] Two different types of struggle were confused here—one "independent of competition," the other conditioned by it. Capitalists and workers resembled the cats and the mice. The more intensely capitalists oppressed workers, the less workers competed among themselves—and so the less they produced. If this oppression were eased, workers would compete more strenuously among themselves, and this would lead to greater well-being.[81] Similarly, if external factors inhibiting the accumulation of capital were relaxed—Bibikov listed "political institutions, luxuries, vanity, extravagance, ambition, greediness"—this would intensify the struggle among capitalists and so generate a more rapid accumulation of wealth.[82] In sum, the relaxation of intergroup struggle would increase intragroup competition and so ensure social progress.

Malthus's law of population, then, held true only in societies that were "not moving forward"; that is, in those countries that were not eliminating obstacles to population growth unrelated to intragroup competition. Inapplicable and class-biased for Europe, Malthus's theory could therefore "render an invaluable service for China."[83]

One is tempted to dismiss Bibikov's essay as a peculiar form of "Social Darwinism." Its peculiarity consisted, however, in his dissection of the Malthusian metaphor and his analysis of the dynamic relationship between its components. This brought him the praise of at least one reviewer and placed him well within the mainstream of Russian reactions to Darwin.[84]

Populism, Natural Science, and the Struggle for Existence

In his review of nineteenth-century Russian literature, S. A. Vengerov indicated that three literary events were the fundamental "signposts" of the emergence of populism. These were the polemic between *The Contemporary* and *Russian Word* in 1864–1865, P. L. Lavrov's *Historical Letters* of 1868–1869, and N. K. Mikhailovskii's essays in *Otechestvennye zapiski* (*Fatherland Notes*) in 1869 and the early 1870s.[85] Each signpost involved a critique of the scientism of the 1860s, with particular emphasis on interpretations of physiological psychology and evolutionism. N. V. Shelgunov never accepted the central dogmas of populism, but agreed by 1879 that he and other columnists for *Russian Word* had overestimated the progressive import of "positive knowledge." "Let us admit," he wrote, "that in the 1860s there were made overly courageous generalizations and conclusions from Darwin; let us agree that we were mistaken [to think] that the [physiologist's] frog would save the world—all this was unfounded and laughable."[86]

The populist criticism of scientism was in large part a response to the transformation of Russian society after the reforms of the 1860s, especially the emancipation of the serfs in 1861. Radicals had previously been preoccupied with the struggle against feudal institutions and ideology, and had found in natural science a natural ally. They were often vague about their ultimate goal, but most hoped that some form of socialism would follow the collapse of tsarism. In the wake of the reforms, radicals became preoccupied with a new enemy; in 1869 N. V. Shelgunov observed that landowners were becoming capitalists and that capital was becoming the driving force in Russian society.[87] Populist commentators in *Fatherland Notes* observed nervously that the social stratification of the peasantry was undermining the peasant commune, which they thought was the basis of Russian socialism.[88]

In his essay "On the Democratic Character of the Natural Sciences" (1875), Mikhailovskii explained the relationship of these changed social circumstances to the political import of science. Preparing an analogy to the 1860s in Russia, he observed that before and during the French Revolution natural science had "served the democratic ideals of equality and freedom" by shattering the cohesive unity of the feudal system.[89] It had undermined Catholic dogma and the entire system of inherited privilege while strengthening an industrial class opposed to the feudal order.[90] After the French Revolution, however—and, for the Russian reader, after the emancipation of the serfs—the nature of political struggles, and the relationship of science to them, changed fundamentally. The same theories that had served democratic ideals in the struggle against feudalism might well prove counterproductive in the new circumstances:

> Darwinism, for example, is democratic precisely insofar as it directly or indirectly undermines those feudal principles which still survive. But not only is it not "democratic in essence," but in the sharpest and most defined manner it places inequality and struggle in society at the cornerstone of its moral-political doctrine.[91]

Populists, then, were predisposed to examine the ideological underpinnings of Darwin's theory and to distinguish those aspects of it that were "friendly to democratic ideals" from those that served their new and frightening antagonist.

Mikhailovskii did so in his essay "The Analogical Method in Social Science" (1869), in which he analyzed the relationship between the laws of nature and society by examining Darwin's struggle for existence. This essay was occasioned by the confession of one Russian columnist, Stronin, that Darwin's theory left him pessimistic about man's future. Mikhailovskii observed that this pessimism was founded on tacit acceptance of Malthus's abstract and discredited doctrine, not on a careful examination of the struggle for existence itself. "The biological law of the struggle for existence is undoubtedly binding for sociology," he observed. But Darwin himself had indicated that this expression encompassed a range of quite varied events:[92]

> Darwin's theory establishes one extraordinarily general fact, specifically, that each individualized entity ... lives, directly or indirectly, at the expense of other organisms or inorganic nature. The direction of the struggle for existence—that is, will it be turned toward a given individual or individuals of the same species, or to individuals of another species, or, finally, to non-living matter—depends entirely upon the particular conditions in which the individual finds itself.[93]

The need to eat indeed required humans to engage daily in the struggle for existence, Mikhailovskii argued, "but Mr. Stronin can obtain his meal by hunting or by agriculture, and in this case his meal will be the result of struggle with individuals of other species or with inorganic nature."[94]

Should humans choose to do so, they could also acquire food through cooperation. Mikhailovskii returned to this point in several subsequent essays. In "Darwin's Theory and the Social Sciences," he termed cooperation in nature "the great fact ... completely ignored by the Darwinists."[95] This omission betrayed Darwinism's roots in utilitarianism, a doctrine that was incapable of convincingly relating individual and societal welfare.[96]

A second leading populist thinker, P. L. Lavrov, addressed Darwin's theory in "Socialism and the Struggle for Existence," which he discussed with Friedrich Engels and published in the radical émigré journal *Vpered (Forward)* in 1873.[97] Living in western European exile, Lavrov rebutted the oft-heard view that Darwin's law of the struggle for existence rendered socialism "a blind fantasy, a contradiction to the laws of nature."[98] He did so by disassembling Darwin's metaphor and examining the historically evolving relationship between its components. "The phases of the struggle for existence vary, it takes different forms," Lavrov wrote. "Finally, it is transformed into a process which resembles the original phenomena of this struggle as little as a mature insect resembles the larva from which it developed."[99]

> Organisms initially engaged in a war of all against all: Families of organisms struggle for control of the soil, ... species of the same family struggle cruelly for the means of life, individuals of one and the same species quarrel intensely over soil, air and food. The casualties are incalculable, nature is full of bodies; of twenty species nine perish, 295 of 357 individuals, four-fifths of the total, lose their lives.[100]

According to Lavrov, individualistic struggle led to the emergence of primitive groups, which enjoyed a competitive advantage against isolated individuals. Intraspecific cooperation among bees, ants, birds, and mammals changed the struggle for existence fundamentally:

> A society of insects struggles to the death with everything surrounding it, but within the society *there is no struggle*, neither between the individual and society, nor between individuals of the society; disregarding danger and its own welfare, each

individual enters the battle for the common goal. The *solidarity* of its members is a tool of the struggle of the *society* for existence.... There exists an *instinctive solidarity, mutual aid* among a defined circle of individuals, and this negation of the struggle for existence between individuals is the most powerful means for the prosperity of their united group.[101]

Lavrov traced the increasing importance of mutual aid in the animal kingdom and linked it to the evolution of human societies.

Capitalists, however, still struggled among themselves; they had not even attained "that instinctive solidarity of like with like that exists among ants." In their efforts to justify the predatory practices of their patrons, bourgeois intellectuals had elevated the lowest phase of the struggle for existence "to a system."[102] "Poor schoolchildren have not noticed that of Darwin's great discoveries they have grasped only one, the very lowest phase of the struggle for existence, and that the animal world, even before man, has already carried this struggle to other, higher phases."[103]

So "the great discoveries of Darwin" did not justify "eternal competition between people and eternal exploitation of some by others." They demonstrated, rather, that "only the solidarity of humanity demanded by socialism can facilitate a future of human development."[104] Engels raised several objections to Lavrov's approach and thought his argument overly moralistic, but allowed that it might be well-suited to the Russian public.[105]

A third leading populist thinker, N. G. Chernyshevskii, was uninterested in reconciling Darwin's theory with socialism. A Lamarckian evolutionist before 1859, he considered the selection theory irrevocably tainted by its Malthusian origin and content. Living in Siberian exile, Chernyshevskii was unable to publish his views until 1888, but he expressed them in letters to his son fifteen years earlier. "Poor Darwin reads Malthus," he explained, "or some Malthusian pamphlet, and animated by the brilliant idea of the 'beneficial results' of hunger and illness, discovers his America: 'organisms are improved by the struggle for life'."[106] Darwin "forgot" that privation always harms organisms, killing some and weakening others. Thus, "the vileness of Malthusianism passed into Darwin's doctrine."[107]

> What is the essence of Darwin's error, and that of his followers? A specialized science, political economy, has undergone such great development (through Ricardo and others, not through Malthus) that it seems capable of providing mathematical truths to science. Darwin noticed this. And made use of what he understood. And did he guess that if you want to use a specialized science for your own work you should study it? No, it seems never to have occurred to him. And the result was the same as if Adam Smith had taken it upon himself to write a course in zoology.[108]

Darwin's Malthusianism remained a "rather innocent folly" in his own work because "concern about the welfare of plants and animals is not an especially important element of our human conscience."[109] But "when this foolishness is transferred to human history it becomes bestial, inhuman."[110] For Chernyshevskii, Zaitsev had captured the logical consequences of the selection theory. If one accepted the arguments of Malthus and Darwin, colonialism and slavery were not only justified but beneficial for human progress.[111]

Chernyshevskii enjoyed great prestige among the liberal and radical intelligentsia. When he published these views under the title "The Origin of the Theory of the Beneficence of Struggle" (1888), even Darwin's most avid supporters took note.[112]

Other Radical Voices: Praising Darwin and Damning Malthus

An influential minority of radical theorists rejected the populists' "subjective sociology" and their distrust of determinist natural science. Gathered largely around the journal *Delo* (*The Cause*), they attempted to remedy the weaknesses in 1860s scientism by combining a determinist view of the natural world with a determinist, but nonreductionist, approach to the social sciences. The Darwin-Malthus connection provided an important arena for this effort since it joined a contemptible political reactionary with the most brilliant naturalist of the age. The general response of these thinkers was that the laws governing the natural and social worlds were fundamentally different.

P. N. Tkachev was an uncompromising critic of Malthus whose "Statistical Essays on Russia" substantiated the familiar argument that repressive social conditions retarded the necessary growth of Russia's population. Reviewing Bibikov's volume in 1869, he damned Malthusianism as a backward and pernicious doctrine. With "crude cynicism" and "a false air of scholarship," Malthus had identified natural law with "the true interests, demands and wishes of all the supporters of bourgeois egoism."[113]

Writing from prison the following year, Tkachev addressed the Darwin-Malthus connection directly in "Science in Poetry, and Poetry in Science." For him Darwin's reliance on Malthus's progressions was only superficially paradoxical:

> When you see a stupid monkey cutting at a stone with a knife and an intelligent man cutting bread with a knife then, of course, you find much in common in the activity of these two beings, so different in their organic development. Both are cutting and both cut with a knife. The entire difference lies in why and what they cut.[114]

Malthus and Darwin both proceeded from "the physiological truth that the life and reproduction of organisms depends upon the conditions of nutrition."[115] In the interests of "the exploiters and shopkeepers," Malthus had applied this truth to political economy and had arrived at "blind absurdities." In the interests of science, Darwin had applied it to organic nature and had arrived at "brilliant generalizations."[116]

The problem, then, lay, not in Darwin's theory itself, but in the false analogy between struggle in the natural and social worlds. In nature struggle was a necessary, "fundamental fact" and the essential cause of organic development.[117] In society it was the artificial product of exploitative economic relations and the source of human misery. Only in "the language of a poet" did the struggle for capital, the battle of landowners and factory owners among themselves and against workers, have anything in common with "the struggle of some mollusk with an ammonite."[118]

This fallacious analogy was used by Malthus and his heirs to justify "any barbarism, any swindle." It rested on a confusion of "parent for child"; that is, it attributed class society to a universal struggle for existence, rather than attributing the struggle for existence among humans to class society itself.[119]

Similar views were expressed by L. I. Mechnikov, a geographer, a follower of the anarchist Mikhail Bakunin, and the brother of zoologist I. I. Mechnikov. In his essays "Colonization in Australia and in America" (1880) and "Prospects for Agrarian Reorganization in England" (1883), Mechnikov contended that production often outstripped population growth and that Malthusianism failed, therefore, to explain human poverty and conflict.[120] In "The Struggle School in Sociology" (1884), he observed that Malthus's theory had lost all credibility until it was unwittingly rehabilitated by Darwin.[121]

For Mechnikov, Darwin's discovery that organisms developed through the struggle for existence constituted "a gigantic step forward."[122] Yet other realms obeyed other laws: the law of gravity governed the inorganic world and the law of cooperation the social.[123] So Darwin's accomplishment as a naturalist was totally separable from the "false Darwinian sociology" created by western European thinkers.[124] For sociologists the problem arose, not from Darwin's view of nature, but from the fallacious Malthusian view that society was "a simple aggregate of living beings capable of producing children, adapting and devouring."[125]

Russia's leading Marxist, G. V. Plekhanov, approached the Darwin-Malthus relationship in a similar fashion. For Plekhanov as for Engels, Darwin was the Karl Marx of natural history.[126] He devoted one chapter of his biography of Chernyshevskii to a refutation of the populist's critique of the selection theory.[127] Darwin had certainly overestimated the evolutionary significance of natural selection, Plekhanov conceded, but had never claimed that the struggle for existence was "beneficial."[128] For Plekhanov, Darwin's uncritical appropriation of Malthus's progressions, and the Malthusian tones in *Descent of Man,* testified to the naturalist's poor grasp of social issues and to his "Manchesterian views." Yet "one cannot hold Darwin responsible for the blunders of the Darwinists" or for the influence of "so-called Darwinism" on the social sciences. Darwin himself had recognized the advantages conferred by cooperation in the struggle for existence, and this supported political views "directly opposite to that extreme individualism" preached by many who invoked his name.[129]

Conservative Criticisms: Malthus, Darwin, and the "British National Type"

Conservative intellectuals initially responded to the *Origin* with muted criticism or mild approval. In 1863 *Russkii vestnik* (*Russian Herald*) published a favorable review written by S. A. Rachinskii, whose Russian translation of the *Origin* appeared the following year.[130] M. I. Vladislavlev praised Darwin for his avoidance of "materialist flourishes," and some theologians praised Darwin for rescuing teleology from the attacks of vulgar materialists. Despite misgivings, the censorship apparatus allowed Darwin's work to appear essentially unhindered.[131] Criticism mounted, however, after the publication of *Descent of Man*. Attacks on Darwin became a staple of political reaction in the 1870s and 1880s.[132]

Conservative critics shared with radicals both a vision of Russian society as a coherent, cooperative whole and a critique of the atomistic nature of western European life. While populists identified Darwin's approach to the struggle for existence with bourgeois political economy, conservatives associated it more generally with the British national character. The obvious cultural content of Darwin's theory provided conservative Slavophiles with a striking illustration of their general argument that science was not a value-free source of positive knowledge and social progress.

N. N. Strakhov was an authoritative critic of radical scientism in the 1860s. He had earned a master's degree in zoology for an essay on the wristbones of mammals, wrote frequently about the natural sciences in the *Journal of the Ministry of Education*, and in 1866 translated Claude Bernard's *Introduction to Experimental Medicine* into Russian.[133] Bernard's attempt to separate scientific medicine from the philosophical quarrels between mechanists and vitalists lent credence to Strakhov's position that progress in the biological sciences did not support the philosophical and political conclusions drawn by Russian radicals.

This distortion of the nature of scientific knowledge was a central theme in his review of the *Origin* in 1862. Entitled "Bad Signs," this essay concentrated on Clemence Royer's portentous introduction to the French edition of Darwin's work. For Strakhov, Darwin had discovered an "internal law" of organic development based on "multiplication, improvement and struggle."[134] Yet Royer's conclusion—that human societies should encourage overpopulation in order to guarantee incessant struggle and progress—demonstrated the dangers implicit in a naive application of Darwin's theory to ethics:[135]

> Really, what a remarkable discovery! What does science tell us? When a family has many children and nothing to eat, Malthus ingenuously took this for a misfortune. Now we see that the more children the better, since the beneficial law of competition can come into play all the more strongly. The weak perish and only those chosen by nature, the best, privileged members, endure the struggle—so progress occurs.[136]

Strakhov thus registered the first Russian objection to "Social Darwinism."[137] Given his view that humans chose ideals and established social laws independent of natural law, he had no difficulty praising Darwin while criticizing "Social Darwinism."

His attitude toward the Darwin-Malthus connection changed in subsequent years. By the early 1880s he had adopted a Slavophile veneration of traditional Russian cultural values and had found in Darwin's theory "the moral stamp of the Englishman." The popularity of Darwinism, he concluded, was not explicable by its logical power but by its resonance with English cultural values. Strakhov explained in *The Struggle with the West in Our Literature* (1882) that this illustrated an important truth about scientific theories in general:

> The motion of the sciences and the revolutions which they undergo do not depend on their internal development but, rather, are conditioned by the influences of some other sphere. The dominance and disappearance of doctrines are governed by a force more powerful than science itself.[138]

Strakhov arrived at this position in collaboration with Darwin's most influential conservative critic, N. Ia. Danilevskii (1822–1885). Arrested as a young man for his participation in the radical Petrushevskii circle, Danilevskii had written a master's essay on the flora of Orlovsk province and, from 1853 to 1857, had participated in K. E. von Baer's expeditions to fishing grounds along the Volga and the Caspian Sea. He then worked for the Department of Agriculture, evaluating means of increasing the production of food, especially fish, in travels throughout Russia in the 1860s, 1870s, and 1880s.

His political views changed fundamentally during the 1860s, and his *Russia and Europe* (1869) became "the Bible of Pan Slavism." Danilevskii wrote passionately of the attempt by individualistic, violent western Europe to dominate Slavic peoples and destroy their culture, and predicted the rise of a superior Slavic civilization, led by Russia. His approach to history owed much to his favorite naturalist, Georges Cuvier. For Danilevskii "cultural-historical types" were distinct and essentially invariable. Like the conditions of existence and the separate organ systems of Cuvier's *embranchements*, the historical experience, culture, philosophy, and science of each "type" were integrally related.[139]

Darwinism typified the scientific ideas generated by the English national type:

> The essential, dominating characteristic of the English national character is love of independence, the all-sided development of the personality, and individualism; which manifests itself in a struggle against all obstacles presented by external nature and other people. Struggle, free competition, is the life of the Englishman: he accepts it with all its consequences, demands it as his right, tolerates no limits upon it.[140]

The Englishman struggled from his days as a schoolchild. Running, swimming, boating—all were competitive sports for him. "He boxes one-on-one, not in a group as our Russians like to spar," founds debating societies for "the struggle of opinions," and even establishes mountain-climbing clubs, "not with the scholarly goal of investigation . . . but solely to allow oneself the satisfaction of overcoming difficulties and dangers . . . in competition with others."[141]

This basic English characteristic was reflected in Hobbes's philosophy of the war of all against all, Smith's faith that the struggle of producers and consumers produced economic harmony, and Darwin's theory that the struggle for existence was the basis of evolution and biological harmony. The onesidedness of each theory reflected that of English life itself.[142]

For more than a decade Danilevskii labored on his massive treatise *Darwinism: A Critical Investigation*. The first volume appeared in 1885, the year of his death. The second, edited by Strakhov, was published in 1889. Many of Danilevskii's arguments against Darwin's theory were compiled from western European sources, but his lengthy chapter on the struggle for existence elaborated his own, and Strakhov's, previously-expressed views:

> Darwinism was a purely English doctrine, with all the particularities of orientation of the English mind, and all the qualities of the English spirit. Practical use and competitive struggle—here are two characteristics . . . that give direction to English life and also to English science. On usefulness and utilitarianism is founded Benthamite ethics, and essentially Spencer's also; on the war of all against all, now

termed the struggle for existence—Hobbes's theory of politics; on competition—the economic theory of Adam Smith and all of that primarily English science, political economy. Malthus applied the very same principle to the problem of population.... Darwin extended both Malthus's partial theory and the general theory of the political economists to the organic world.[143]

The cultural roots of the selection theory explained its simultaneous discovery by Darwin and Wallace. Danilevskii also indicated another peculiarly English influence upon it: "The curiosity of English horticulturalists and breeders of domestic animals, leading to exhibitions, races and other competitions between animals," and so to the artificial selection of plants and animals "for the satisfaction of their capricious fancies."[144]

That struggle occurred in nature was indisputable, but Darwin exaggerated its scope and significance. One reason was his mistaken belief that organisms were packed tightly into every available space. For Danilevskii nature contained "wide gaps, voids so to speak, which various animals and plants can fill over a long time without any struggle for existence."[145] Nature did not resemble a single fountain at which a multitude of organisms battled to drink, but rather the shore of a long river, where crowding at one drinking spot merely required an organism to travel to another.[146]

According to this view, which flowed easily from Danilevskii's travels throughout Russia, a dearth of food need not lead to incessant struggle. Population might outstrip food supply in one location, leading to "temporary and local wars," but this struggle would be neither sufficiently continuous nor intense to lead to natural selection.[147] Furthermore, given the multiplicity of changing and often contradictory environmental pressures, struggle could not lead to directional development.

Darwin had contradicted his general thesis by asserting that expansive, complex environments were most conducive to the evolution of new forms. For example, Darwin explained the stability of the rabbits of Porto-Santo as the result of their isolated habitat. Yet if Darwin's analysis of the struggle for existence were correct, isolation should stimulate evolution:

> A many-sided struggle, such as occurs in the large areas of land and sea, by virtue of the variety of competing [environmental] elements, prevents any one of them from attaining dominance for a long time, and consequently cannot facilitate the adaptation of organisms to it. The intensity of struggle is interrupted, or acquires another direction, and this preserves existing forms.... But the opposite should occur in limited locales, where, by their very one-sidedness, the less varied elements of struggle are not counterbalanced. There one [environmental] condition acquires constant dominance, and the process of struggle and competition among individuals begins in relation to it—that is, specifically that form of struggle which should have led to the accumulation of useful variations, to selection.[148]

The stability of isolated species, such as the rabbits of Porto-Santo, cast doubt on Darwin's basic postulates. The constantly changing and often contradictory demands of large and complex environments conferred a complex set of advantages and disadvantages on different, often competing organisms. The result was that struggle tended to preserve a "balance in the numerical strength of species" and was "a condition that preserves more than it alters, acting more as a conservative than a progressive force."[149]

Nor did an increasing population usually generate struggle and directional change. It resulted most frequently in the more rapid circulation of natural resources, just as increases in human populations resulted in the accelerated circulation of capital. The wide gaps in nature, and the rapidity with which its resources could conceivably circulate, revealed that the earth could support a much larger population without generating a Darwinian struggle for existence.

Since Danilevskii's volume appeared posthumously, Strakhov became its chief publicist and defender. This role embroiled him in a heated polemic with the orthodox Darwinian K. A. Timiriazev from 1887 to 1889.[150] Anxious to move on to other interests, Strakhov noted contentedly in a letter to Tolstoy that he had accomplished his task: "Everybody is talking about Danilevskii's *Darwinism*."[151]

Religious thinkers focused on the struggle for existence in their comments about Darwin's theory during the last third of the nineteenth century. K. P. Pobedonostsev captured their moral objection to Darwin in his comment that "it is plain that to him the fundamental law of life is the preservation of the strong and the expiration of the weak."[152] Some, such as N. Rumiantsev (1895–1896), repeated Danilevskii's arguments.[153] Others, such as M. Glubokovskii (1892), rejected the ruinous consequences on morality of Darwin's emphasis on intraspecific conflict and found a necessary corrective in the theory of mutual aid.[154]

Tolstoy on Malthus and Darwin

Russia's most celebrated novelist of the nineteenth century despised Malthus as a malicious mediocrity who sought to justify poverty, competition, and individualism.

> A very poor English writer, whose works are all forgotten, and recognized as the most insignificant of the insignificant, writes a treatise on population, in which he devises a fictitious law concerning the increase of population disproportionate to the means of subsistence. This fictitious law, this writer encompasses with mathematical formulae founded on nothing whatever; and then he launches it on the world. From the frivolity and the stupidity of his hypothesis one would suppose that it would not attract the attention of anyone, and that it would sink into oblivion, like all the works of the same author which followed it; but it turned out otherwise. The hack-writer who penned this treatise instantly becomes a scientific authority, and maintains himself upon that height for nearly half a century.[155]

For Tolstoy, the enduring popularity of Malthus's theory attested to the influence of prejudice on science. Weak and erring folk simply used "the imposing word 'science'" to sanctify their views.[156] The practical consequences of Malthusianism illuminated its origins, not in abstract "science," but in the instincts of a corrupt intelligentsia.

> The deductions directly arising from this theory were the following: the wretched condition of the laboring classes was such in accordance with an unalterable law which does not depend upon men; and, if anyone is to blame in this matter, it is the hungry laboring classes themselves. Why are they such fools as to give birth to children, when they know that there will be nothing for them to eat? And so this deduction, which is valuable for the herd of idle people, has had this result: that

all learned men overlooked the incorrectness, the utter arbitrariness of these deductions and their insusceptibility to proof; and the throngs of cultivated, that is, of idle people, saluted this theory with enthusiasm, conferred upon it the stamp of truth, that is, of science, and dragged it about with them for half a century.[157]

Darwin had introduced these same values into biology.[158] His theory illustrated the manner in which science tendentiously selected its facts and interpretations. By extending Malthus's views to nature, Darwin had further justified human "idleness and cruelty."[159] Tolstoy praised Strakhov, Danilevskii, and Chernyshevskii for their exposure of Darwin's Malthusianism.[160] In *Anna Karenina* his character Levin attacked the moral consequences of Darwin's views so sharply that Timiriazev felt compelled to respond.[161]

Tolstoy returned to this theme repeatedly, perhaps most dramatically in a final letter to his children, dictated from his deathbed in 1910 and warning that "the views you have acquired about Darwinism, evolution and the struggle for existence will not explain to you the meaning of your life nor will they provide guidance in your actions."[162]

Conclusion

We have seen that Russian social thinkers of the Left, Right, and center shared anti-Malthusian sentiments, identified Darwin's struggle for existence with Malthus, and reacted critically to this Darwin-Malthus connection. This common reaction was refracted through the prism of varied ideological perspectives and political agendas. Strakhov and Danilevskii used the obvious tendentiousness of Darwin's doctrine to illustrate the cultural subjectivity of scientific theories and to warn Russians against the importation of alien Western values in the guise of positive knowledge. Pisarev and Zaitsev sought to incorporate Darwin's theory into radical scientism, but the price, an apparent softening of the Left's opposition to slavery and colonialism, was unacceptable to most of their comrades. Nozhin and Chernyshevskii attempted to preserve an evolutionism free of bourgeois ideology by rejecting Darwin's theory as inherently Malthusian. Bibikov, Lavrov, and Mikhailovskii dismantled Darwin's struggle for existence and reassembled its component parts in a manner congenial to their political views. Tkachev, Mechnikov, and Plekhanov insisted that the problem lay, not in Darwin's Malthusianism, but in the superficial analogy between biology and society; their critique of the struggle for existence enabled them to discard Malthus while praising Darwin.

It would indeed be surprising if scientists had not reacted similarly—if they had not been suspicious of a metaphor so foreign to their culture and so laden with negative associations, if they had not explored the possibility of separating the great Darwin from the unsavory Malthus. Many leading scientists did just that.

CHAPTER 3

Beketov, Botany, and the Harmony of Nature

> Life competition leads to balance rather than unending extermination. This, I think, incontrovertible conclusion is completely opposed to those blind, I will even say morally repugnant, conclusions that Malthus drew from his false law and which even today many take for honest coin.[1] (A. N. Beketov, 1891)
>
> An ardent Westernizer in everything,
> At heart he is an old Russian lord.[2] (Alexander Blok, on Beketov)

Although few recognize his name today, in the second half of the nineteenth century Andrei Nikolaevich Beketov (1825–1902) was Russia's premier botanist and the generally acknowledged "Father of Russian Botany."[3] During these decades of rapid growth for Russia's scientific community, Beketov wielded great institutional authority, trained a generation of influential students, and was a widely read popularizer of science. He enjoyed warm relations with many of Russia's other famous figures, including the author Fyodor Dostoyevski, the discoverer of the periodic table D. I. Mendeleyev, and his own grandson, the poet Alexander Blok. As one of Russia's many evolutionists before 1859, Beketov read the *Origin* with great interest and struggled throughout his life to reconcile the selection theory with his own views. The chief difficulty here stemmed from the tension between Darwin's concept of the struggle for existence, with its Malthusian flavor, and Beketov's notion of dynamic harmony, which combined his deeply held religious beliefs and a mechanistic view of nature into a Russian variant of evolutionary natural theology. This tension, and Beketov's resolution of it, reveal that while Russia's transformist tradition encouraged acceptance of Darwin's evolutionism, it also generated resistance to the novel aspects of his selection theory.

Russia's Premier Botanist

Beketov was born in 1825 to a wealthy gentry family of Penzensk province in central Russia. His father, Nikolai Alekseevich, was a well-read man with a literary bent whose circle included the well-known poet E. A. Baratynskii (1800-1844). His mother died when he was four. The four children were raised by a Swiss nanny, tutored in religion, mathematics, and grammar, and sent to St.Petersburg for further education. Ekaterina studied at Smol'nyi Institute and returned to the family estate. Aleksei pursued a career in engineering. Nikolai became a renowned professor of chemistry and a member of the Academy of Sciences.

Andrei's studies began inauspiciously. After a lackluster year in the department of Eastern languages at St. Petersburg University, he willingly surrendered his ambition to become a philologist and accepted his father's suggestion that he transfer to the Aleksandrov Military Academy. But he proved an equally uninspired and inept cadet, and so embarrassed his company during a review by the tsar that he was encouraged to leave.

St. Petersburg, however, offered other interests. Like many youths in the capital the three Beketov brothers were attracted to Fourier's socialist ideas, and they formed one of the many circles that sought in Fourierism a modern worldview.[4] One member of their circle was Fyodor Dostoyevski, who lived with them for about two years while writing his first novel, *Bednye liudi* (*Poor People*). In a letter of 1846 he wrote effusively of the intelligence, warmth, and noble character of "my good friends the Beketovs."[5]

In 1846 Andrei left St. Petersburg to enroll in the physicomathematical faculty of Kazan University. Nikolai followed one year later. Without them their circle soon disintegrated; several of its members, including Dostoyevski, joined a similar group led by M. V. Petrashevskii. They became famous as the "Petrashevskii circle" when they were arrested for revolutionary activity in the summer of 1849.

V. V. Bervi-Florinskii (1829-1918), who became a leading socialist thinker, later recalled the impact of the Beketov brothers on student culture in Kazan:

> Like Petrashevskii they spread Fourier's teachings, and the results were the same as in St. Petersburg. They quickly acquired great influence in the university. The newcomers from St. Petersburg taught us to study science independently. Until then Kazan University students did not have the slightest notion that they could study science rather than memorize lectures.[6]

There were at the time few students of natural science at Kazan University. Aside from the Beketov brothers, the close-knit group included N. P. Vagner (1829-1907), later a professor of zoology at Moscow University, and A. M. Butlerov (1828-1886), who became a professor of chemistry at Kazan and St. Petersburg universities and a member of the Academy of Sciences.[7]

Although committed to science studies, Andrei was unimpressed by his teachers and found "neither mentor nor patron" at Kazan University.[8] Typical of the lackluster faculty, he later recalled, was professor of zoology E. A. Eversmann, who spoke Russian poorly and whose lectures consisted chiefly of halting translations from a German textbook of systematics.

After graduation in 1849 Beketov traveled to Tbilisi, Georgia, where he taught for several years at a gymnasium. He instructed students in natural science, physics, mathematics, and agriculture until the arrival of new faculty members allowed him to specialize in natural science. He became especially interested in botany and studied the regional flora in numerous expeditions from 1850 to 1855. One product of these travels was his *Essay on the Flora of Tbilisi*, for which St. Petersburg University granted him a master's degree in 1853.[9]

While in Tbilisi he befriended Alexis Collins, a devout Catholic and ascetic who worked at the French consulate. The two walked for days in the wilderness, discussing Plato, Leibniz, and Christian morality. Beketov described the peaceful pleasures of these retreats in "Reminiscences of Tbilisi and Its Environs," which was published in the journal of the Imperial Russian Geographical Society in 1855.[10] His abiding sense of harmony in nature, and the quiet religiosity and tendency toward asceticism often noted by later acquaintances dated from this period.

During one trip on horseback between Tbilisi and St. Petersburg, he met the daughter of the famous naturalist G. S. Karelin (1801–1872). The two were married in 1854 and took up residence on the Karelin estate near Moscow. Beketov used Karelin's rich botanical collection for many subsequent works, including his doctoral thesis, *On the Morphological Relations among the Parts of Leaves, and between the Leaf and the Stem* (1858).[11]

By this time his father, like many gentry of the 1850s and 1860s, had lost his fortune, and for Beketov, as for many children of the declining gentry, science became a means of support. In 1859 he was appointed professor of botany and director of the botanical garden at Kazan University. The salary was low and the climate taxed his wife's health, so in 1860 they moved to St. Petersburg, where the university's Department of Botany was vacant. Beketov served as an occasional lecturer until 1863, when he was appointed professor of botany.

He supplemented his salary by selling articles to popular journals and by translating foreign scientific works into Russian. Among his contributions to M. N. Katkov's popular journal *Russian Herald* were "Zootomia" (1857), "The Northern Ural Region" (1857), "Renewal and Metamorphosis in the Plant World" (1858), "Essays on Virgin Nature" (1858), "The Climate of European Russia" (1858), and "Harmony in Nature" (1860). His *Conversations about the Earth and the Creatures Living upon It* ran through eight editions between 1864 and 1903 and sold some fifty thousand copies, a most impressive total for tsarist times.[12] By the end of the 1870s he had also translated and edited eleven books, including T. H. Huxley's *Man's Place in Nature* (1864), Darwin's *Voyage of the Beagle* (1865), A. R. Wallace's *Malay Archipelago* (1872), and several botanical works.[13] The Darwin volume may well have been a collaborative effort with his wife, Elizaveta, a talented and erudite woman whose friends included Dostoyevski, Tolstoy, and Gogol, and whose long list of translations included H. T. Buckle's *History of Civilization in England* and several novels by Louisa May Alcott.

During his more than three decades at St. Petersburg's Department of Botany, Beketov became Russia's most prominent botanist, a leading figure in the organization of Russian science, and a revered and well-connected member of the liberal intelligentsia. His contemporaries hailed him as the "Father of Russian Botany" and had ample grounds to do so, for Beketov did much to provide Russian botany with the institutional, pedagogical, and scholarly apparatus of a successful discipline. He chaired the nation's premier department of botany from 1863 until 1884, served as director of its botanical garden, and, in 1886, founded and coedited *Botanicheskie zapiski* (*Botanical Notes*), the country's first Russian-language botanical journal. He developed a Russian botanical lexicon and wrote three widely-used university textbooks: *A Botanical Course for University Students* (1862–1864, 1871), *Textbook of Botany* (1880–1883, 1885, 1897), and *The Geography of Plants* (1896). Beketov also devised a curriculum for botanical studies in gymnasia that was endorsed and widely publicized by the Ministry of Education.[14]

Yet Beketov's greatest contribution to Russian botany may well have been his students, who dominated the discipline well into the Soviet period. Three of the most important were K. A. Timiriazev (1843–1920), Darwin's most famous Russian disciple and a professor at Moscow University and the Petrovsk Academy of Agriculture and Forestry; A. S. Famintsyn (1835–1918), professor of botany at St. Petersburg University and founder of Russia's first laboratory of plant physiology at the Academy of Sciences; and V. L. Komarov (1869–1945), chair of the Department of Botany at St. Petersburg University and president of the USSR Academy of Sciences from 1936 to 1945. By the end of the century these and other Beketov students populated the faculty of every major Russian university.[15] Beketov's pedagogical influence extended even to the Winter Palace, where for two years he tutored Prince Paul Alekseevich.

In his scientific work Beketov maintained his early interest in plant morphology but, by the late 1860s, concentrated on botanical geography. Over the previous century P. S. Pallas, I. G. Gmelin, R. E. Trautfetter, and I. G. Borshchev had described much of Russia's flora. Beketov drew on this material, the work of his students and colleagues, and his own investigations of the St. Petersburg and Moscow regions and the steppes of southern Russia to delineate the country's botanical zones and to develop a theoretical understanding of the relationship between flora and the physical environment. His approach owed much to his incorporation and critique of the ideas of Alphonse de Candolle and August Grisebach.[16] Such works as "On the Flora of Archangel" (1884), "The Phytogeography of European Russia" (1884), "On the Flora of Ekaterinoslav" (1886), and *The Geography of Plants* (1896) testified to Beketov's conviction that climate and topography governed the distribution of plants.[17]

His preoccupation with unifying generalizations is evident from one student's recollection that he lectured at such length about philosophy and evolutionary theory that the semester often ended without mention of systematics.[18] He shared his passion for these subjects in such essays as "On the Struggle for Existence in the Organic World" (1873), "Do We Find Disharmony in Nature?"(1876), "Human Nutrition at Present and in the Future" (1878), "Darwinism from the Perspective of the General Physical Sciences" (1882), and "Morality and Natural Science" (1891), and in his short story "Doctor Froman" (1892).[19]

Beketov also played an important role in the organization of Russian science as a whole. Together with K. F. Kessler he organized the First Congress of Russian Naturalists and Physicians in 1868, and he presided over the sixth congress in 1879 and the eighth in 1890. He served as secretary of the St. Petersburg Society of Naturalists and editor of its journal from 1869 until 1881, when he succeeded Kessler as the society's president. He was also secretary of the Imperial Free Economic Society, the author of its official history, and the editor of its journal from 1889 to 1897.[20] The society's goal of putting science to practical use found expression in

Beketov's many articles on soil conditions, crop cultivation, and the wine industry.[21]

A popular member of the St. Petersburg University faculty, he served as dean of the physicomathematical faculty from 1867 to 1876 and was the university's last faculty-elected rector in the turbulent years 1876–1883. As rector he clashed constantly with higher authorities. However faded his youthful radicalism, Beketov proved a determined defender of university autonomy and of arrested students and faculty. In his autobiography, written in the third person, he later recalled that his efforts were "rewarded, at least in his first years as rector, by good relations with the students, which allowed him to calm disturbances without any, or almost any, youthful victims."[22] One liberal observer described him in 1884 as "a remarkable scholar, a very kind and good man," but noted that "opinions varied on his performance as rector and state official: those above said 'he is weak' while those below thought him 'flexible'."[23]

Beketov was particularly proud of his role in the founding and administration of the famous Higher Women's Courses. Women had been largely excluded from Russian universities until, in 1876, four hundred women petitioned Kessler, then rector, to establish a system of public lectures for them. Kessler turned to Beketov, who, together with N. V. Stasova, established an increasingly formalized curriculum featuring such leading St. Petersburg academics as Butlerov, Mendeleyev, Famintsyn, and I. M. Sechenov. This school became known as the "Bestuzhev Courses" in honor of its first official director, a conservative academic. Beketov's daughter later claimed that "Beketov Courses" would have been a more accurate label, but that this honor was denied her father due to his "reputation as a Robespierre" among tsarist officials.[24]

The University Statute of 1884 tightened state control over universities, and the authorities immediately used its provisions to move against Beketov. Minister of Education I. D. Delianov chose a new rector and appointed Beketov's former student, the aristocratic and politically conservative Kh. Ia. Gobi, as the new chairman of the Department of Botany. In 1888 Beketov was removed from state service; only repeated faculty requests preserved his right to lecture at the university, which he did, at a greatly reduced salary, until 1897.[25] He and Stasova were purged from the Higher Women's Courses in 1889.

An enthusiastic participant in the life of the liberal St. Petersburg intelligentsia, Beketov, as rector, enjoyed a healthy salary and spacious living quarters, which permitted him and his wife to host large Saturday evening gatherings of professors and students. Among the regular guests was Mendeleyev, who persuaded Beketov to purchase an adjoining estate, Shakhmatova.[26] The two became fast friends. Beketov's daughters attended the Women's Higher Courses, worked for the liberal journal *European Herald*, wrote poetry, and translated foreign literature.

One daughter, Alexandra, contracted a short, unhappy marriage to a young law student, A. L. Blok. She returned, pregnant, to Shakhmatova and raised her son there. The son, Alexander, fell in love with Mendeleyev's daughter Liubov, married her in 1903, and immortalized her in his "Stikhi o Prekrasnoi Dame" ("Verses about a Beautiful Lady"). These romantic poems were among Alexander Blok's earliest creations.

Beketov suffered a stroke in 1897 and lay paralyzed for five years until his death in 1902. The St. Petersburg Society of Naturalists devoted a volume to the affectionate reminiscences of his students. All praised him as a scholar, teacher, popularizer, and organizer of Russian science. The terms in which they did so, however, revealed differences among the generation of botanists that he had trained. Neovitalist I. P. Borodin pointed to the hymn with which Beketov had introduced *The Geography of Plants* and emphasized his mentor's abiding religious faith. V. L. Komarov recalled that Beketov's belief in the compatibility of religion and science had left him all the more appalled by idealist views that compromised the independence of scientific inquiry.[27]

Alexander Blok remembered his grandfather with a poem:

> His death or sleep was what we all expected.
> The moments passed, weary, interminable.
> A sudden breeze came through the open window,
> Stirring the pages of the Holy Bible.
>
> Out there, a man with snow-white hair was walking.
> Gait brisk and eyes that brimmed with merriment;
> He smiled at us and beckoned with his finger,
> And with familiar steps away he went.
>
> And suddenly, those of us who were there,
> Both young and old, all came to realise
> Who'd been in front of us; we turned, trembling,
> To find the lifeless clay there with closed eyes . . . .
>
> But it was sweet to look on such a soul
> And see its merriment as it went away.
> Our time had come to love and remember,
> And mark a different kind of moving day.[28]

Harmony and Evolution

Months before word of Darwin's *Origin of Species* reached Russia, Beketov completed a lengthy essay about natural relations and evolution, which was published in the *Russian Herald* in 1860. The essay's title, "Harmony in Nature," revealed the metaphor most important to his thinking.

For Beketov the word "harmony" denoted the mutually dependent relations among organisms and between organisms and physical conditions. "The harmony of nature," he explained, "is a manifestation of the law of universal necessity, and its essence is the mutual dependence of all the material parts and phenomena of nature."[29] People sensed this universal interconnectedness and derived great comfort from it. They might lose this feeling amid the bustle of urban life but could regain it quickly by a walk in the wilderness. Science sought the "laws of universal harmony" that underlay this feeling and governed the natural world.[30]

This notion of natural harmony owed much to the French zoologist Georges Cuvier, whom Beketov greatly admired. Surveying the plant and animal kingdoms, Beketov illustrated the perfect fit between the parts of an organism, and between the coloration, structure, and physiological functions of living beings and their physical conditions of existence.[31]

He departed from Cuvier, however, in his interpretation of these harmonious relations. One could attribute them, he explained, either to design or to a natural process of dynamic mutual dependence. Why, for instance, did the watermelon have such a weak stem and bear its heavy fruit on the ground? Either "nature, foreseeing heavy fruit" provided a weak stem so the fruit could be supported on the earth or the watermelon's weak stem led to the development of its weighty product.[32] Similarly, how was one to explain a heavy cross on the roof of a church? One might, by the teleological approach, assume that the church had been constructed to support the cross. Conversely, one might assume that the cross was chosen because its weight and dimensions suited the church it was to adorn.[33]

Rejecting teleology, Beketov contended that harmonious relations resulted from "the adaptation of every phenomenon to its specific purpose and to the environment of its activity."[34] God had invested matter with specific physical qualities, including the capacity to change form constantly in accordance with physical laws.[35] Like molten metal poured into a mold, every material entity was shaped by the particular conditions in which it appeared. But life was distinguished from nonlife by its quality of continual self-renewal. Therefore:

> If the conditions surrounding one or another of these beings change then it must itself undergo a fundamental transformation or be destroyed.... Consequently we notice two phenomena: (1) the variability of beings to the degree that the conditions surrounding them change, and (2) their complete disappearance with a radical change in these conditions.[36]

The fossil record demonstrated that changing physical conditions had caused the extinction of some forms, such as the dodo, and the emergence of new ones. In recent times humans had become a major agent of change, eliminating wolves from the British Isles and pushing the Australian kangaroo to the brink of extinction.

A most striking harmonious relationship in the plant and animal kingdoms was that between life conditions and the rate of reproduction. Like Malthus, Paley, and Darwin, Beketov was struck by the capacity of organisms to multiply rapidly. For instance, he wrote of the opium poppy, *Papaver somniferum*, "In four years its seeds could give rise to an unbelievable number (1,025,000,000) of offspring, so if they all developed into *Papaver somnifera* they would cover the entire surface of the earth."[37] All freshwater plants shared this capacity for "infinite multiplication," as did some animal forms.[38]

The varying fecundity of organisms conformed to a natural law: the greater the environmental dangers to which an organism was exposed, the greater its means of reproduction. Beketov devoted many pages to this "law of harmony":

> It is well known, for example, in what quantity herring are exterminated, not only by man but also by various other animals, and this little fish is so fertile that if only a single one existed in all the oceans, and if it were possible to guarantee the preservation of its offspring, then in several years its progeny would fill all the earth's oceans from their surface to their deepest depths.

Large and predatory fish are not so fertile, and still less so are the scaled reptiles, while the innocent frog has an enormous quantity of eggs.

The least fertile birds are, again, the large and predatory ones, and the most are the defenseless chickens and small birds.[39]

For Beketov the conditions surrounding an organism constantly regulated its fertility, which therefore varied as necessary to preserve its existence. If herring were not constantly exterminated in such large numbers, they would not reproduce so rapidly and so would never, in fact, "fill all the earth's oceans from their surface to their deepest depths."

Unless physical conditions changed very rapidly, as they had in the distant past, the relations among organisms usually led not to the complete displacement of any form, but to balance. Beketov cited the well-known case of the introduction of goats to the island of Juan Fernández. The goats reproduced rapidly, attracting pirates with the prospect of a good meal. To discourage these visits the inhabitants brought dogs to the island. The dogs killed many goats and soon rivaled them in number. The remaining goats "became very careful" and confined themselves to areas that were relatively inaccessible to the dogs. Deprived of a reliable supply of meat, the dogs slowly decreased in number until "a balance was established both between the goats and the dogs and between the goats and the plants." A precise, scientific understanding of this balance would require an analysis of the soil and climatic conditions of the island.[40]

Beketov summarized his central proposition as follows: "The structure, external appearance and the entire essence of each being is caused by the surrounding conditions and dependence upon these conditions—in short, by harmony."[41] The simple recognition of harmony, however, explained nothing. It remained "to examine thoroughly all the conditions of existence of each being; only then will we discover how these conditions, and the totality of their effects," generated specific organisms and defined the relationships between them.[42]

For Beketov, then, natural relations could be summarized by the metaphor "harmony." This metaphor certainly expressed his faith in the ultimately benign purposefulness of nature, yet he explicitly rejected teleological explanations of that purposefulness. "Harmony" was a summary of natural relations and a guide to studying them—not a substitute for scientific investigation itself. By the time "Harmony in Nature" was published, Darwin had offered a radically different metaphorical shorthand for the natural relations resulting in evolution. The tension between Darwin's "struggle for existence" and Beketov's "harmony of nature" would preoccupy Russia's leading botanist for the rest of his life.

Harmony and Struggle, 1864–1876

Beketov's initial comments about Darwin's theory were generally sympathetic and revealed his effort to reconcile these two different metaphors. Although acquainted with the *Origin* by 1864, Beketov did not mention Darwin in "Two Public Lectures on Acclimatization" (1864) and "Are Plant Forms Adapted to Light?"(1865).[43] His approach to evolution remained unchanged from 1860:

We will not say, as was said in earlier times, that everything in nature is mutually adapted in the best possible way, in accordance with certain goals. Rather, we hold that everything in nature tends to take a form that corresponds completely to surrounding conditions. ... In this constant tendency of beings toward mutual adaptation, toward achievement of an organization most in agreement with their life goals, naturalists should seek the causes of the forms assumed by numerous organisms and of the constant improvements which these forms undergo over the centuries.[44]

Darwin's influence was, however, evident in Beketov's use of the word *struggle* (*bor'ba*) for the first time in his published work. Beketov, however, referred only to "struggle" between organisms and physical conditions. Plants waged a "constant and stubborn struggle with the elemental forces of nature at the polar and mountainous boundaries of their diffusion"; the northernmost boundaries of plant growth in the tundra presented a "theatre of struggle," plants had "struggled for decades" with the snowstorms of the tundra, and so forth.[45]

Beketov first commented publicly about the selection theory in an essay whose title revealed his central concern: "The Struggle for Existence in the Organic World" (1873). Here he praised Darwin fulsomely but insisted that the "brilliant English naturalist" had only begun the necessary analysis of the dynamics of the struggle for existence.[46]

This expression, Beketov observed, offered only an imprecise characterization of relations in nature. Many different phenomena could, after all, be labeled a "struggle for existence": one could speak of the earth's "struggle for existence" against the gravitational pull of the sun, a crystal's against the erosive effects of a solution, or the crust of the earth's against the internal terrestrial forces pressing upon it. Clearly, the "struggle for existence" was simply a "foreign expression" for the interrelationship of physical forces.[47]

Darwin deserved great credit for directing naturalists' attention to the evolutionary significance of the struggle for existence—that is, of the complex of relations in the natural world. But this in itself explained nothing. "If we are satisfied with the explanation that every form and every adaptation is necessarily due to the struggle for existence, then we will have contented ourselves with mere words."[48]

When a plant is introduced into a new habitat, Beketov asked, "with what conditions must it struggle?"[49] That depended on the circumstances. If physical conditions in the new habitat differed markedly from those in the old, then the plant's direct relationship to those conditions usually proved decisive. Frequently, however, the newcomer's "most dangerous enemies" proved to be other, better-adapted plants. This indirect competition among organisms "is the struggle upon which Darwinism places special emphasis," and so deserved special attention.[50]

Struggle among organisms was multidimensional. For one thing, organisms struggled both for their individual survival and for that of the group. "The organs of nutrition serve the plant as a tool of the individual's struggle for existence; the organs of reproduction are tools for the struggle of the entire species."[51] In the animal kingdom, also, "we see this dual struggle: ... the struggle for preservation of the individual and the struggle for preservation of the species."[52] Species possessing greater means of struggle for the preservation of the group (that is, rapid reproduction) possessed less for preservation of the individual (that is, means of acquiring nutrition). High fertility sometimes stimulated individual struggle, but it was also "the most powerful tool of [a species] struggle against harmful environmental conditions."[53] Turning to indirect struggle among plants and animals, Beketov offered several examples supporting Darwin's contention that the intensity of such conflict was proportional to the similarities between them, "so individuals of the same species are engaged in the most intense struggle among themselves."[54] He also agreed that this conflict "leads to the development of more complex forms, as Darwinism demonstrates."[55] The swiftness of antelopes, for example, had clearly evolved through such indirect competition. Slower individuals had fallen victim to predators, while swifter ones survived and multiplied. Stopping just short of endorsing the selection theory, Beketov observed that in such cases "the struggle for existence, in the comparatively narrow sense in which Darwinists accept it, occupies the primary place in the theory of the formation of species by means of natural selection."[56]

Yet this "very same logic" led Darwinists to other, paradoxical conclusions. For instance, if similar life requirements always increased the intensity of the struggle for existence the most intense struggle should ensue between the parts of a single organism. Yet

> the shoots of a tree find themselves in the same antagonism, in the same kind of struggle, as occurs between the arms of a balance; eliminate one of the arms and the other will become victorious, but the balance thereby . . . ceases to be a balance. A shoot that grows more energetically than the others, and at their expense, harms the entire tree, and . . . can cause the death of the entire complex organism. This illustrates the paradoxical nature of conflict; here, obviously, it results in the preservation of balance and not in the death of one or another part.[57]

Similarly, one could speak of the struggle between the left and right arms of a human being. "Logically, the expression 'struggle' is completely applicable here in the Darwinist sense, and its strangeness and paradoxicality are fully evident."[58] One could insist that when the left arm is destroyed the right receives more nutriment and grows larger, but this was clearly absurd. "The physiological antagonism of the parts has as a normal result not the death of one part of the whole, but rather the balance of these parts."[59]

The limitations of the Darwinists' concept of the struggle for existence were also illustrated by the prevalence of mutual aid. Antagonisms among organisms were often diminished by their common conflict with physical conditions. "Sympathy between mammals or birds living socially is so great that it leads sometimes to self-sacrifice; it is a powerful influence supporting and strengthening a tendency to sociability and it is a powerful means of self-defense."[60]

By virtue of its dependence on Malthus, Darwinist analysis was especially weak with respect to humans. Beketov repudiated Malthus's "depressing idea" that "humanity would perish sooner or later from lack of food" and that struggle among humans would inevitably become increasingly severe.[61] Malthusians believed that the intensity of the struggle among people was directly proportional to the closeness of their relationship; civil wars were more destructive than those between different peoples, and still more severe was the indirect competition among inhabitants of crowded villages and cities. Yet this logic led inescapably to the conclusion that struggle was most intense within a family. One need only consider the "picture of a father and his beloved son driven by thirst to a bloody battle over a drop of water" to see the bankruptcy of such reasoning and the fallacy of characterizing all relations as "struggle."[62]

Malthus also ignored the human capacity to change with changing conditions and so to avoid the consequences of his abstract law.[63] Human evolution featured the transformation of conflict into mutual sympathy and Christian love, which was "an expression of a quality naturally inherent to man and developed over many thousands of years under the influence of surrounding conditions." This quality played an increasingly important role in social relations and would eventually lead to the elimination of inequality, poverty, and, therefore, human conflict itself.[64]

Beketov's two main conclusions reflected his effort to reconcile Darwin's struggle for existence with his own concept of harmony in nature. He observed, first, that "balance is the goal (in the sense of result) of the so-called struggle for existence."[65] Organisms were constantly adapting to environmental circumstances. Their numbers increased and decreased; their means of reproduction, nutrition, and defense constantly changed—regulated, in part, by the struggle for existence. But this struggle rarely led to extinction or the complete displacement of one species by another. Much more frequently, it culminated in a new balance, perhaps with one organism more dominant than it had been previously.[66]

Second, he praised Darwin and Wallace for demonstrating the evolutionary role of the struggle for existence, but warned naturalists not to substitute this metaphor for a more rigorous, mechanical appraisal of organic relations. Beketov insisted, as he had earlier with respect to his own metaphor of harmony, that the struggle for existence must ultimately be understood as "simply the interaction of general physical forces at greater or lesser complexity."[67] Only by demystifying the struggle for existence—by breaking it down into its component parts and understanding it in terms of physical laws—could science develop the ideas "at the foundation of the great theories of Lamarck and Darwin."[68]

Three years later Beketov again addressed his central concern about Darwin in an article for the popular journal *Priroda* (*Nature*), "Do We Find Disharmony in Nature?"(1876).[69] Here he took issue with such authors as I. I. Mechnikov, who interpreted imbalances in human development and the death of organisms through environmental catastrophe and illness as evidence of natural disharmony.[70] The essence of such views, Beketov commented, was the equation of disharmony with "the incomplete adaptation of organisms to the environment, and to the material goals of their existence."[71] Yet the structure of all organisms was determined by the conditions in which they arose, and so was ultimately explicable by physical laws. "Every being, having taken shape from specific matter in specific conditions, is complete in the sphere of these conditions."[72] Whatever problems pregnancy presented to a woman, the fetus could not be carried to term in any other way and so this physiological process "is in complete harmony with the environment," as surely as a mathematical sum was dictated by its separate parts.[73]

Horticulturalists and breeders shaped nature for human ends, but their work hardly constituted a correction of natural disharmonies. Trees without thorns, fruits without seeds, and cattle of sickly obesity would all vanish rapidly under natural conditions. To see these forms as improvements on nature led to such "colossal stupidities" as the Malthusian doctrine:

> We would have to recognize as a correction of nature the castration of cows, horses and other animals, and even of man. And, after all, the Skoptsy [a religious sect that practiced self-castration] consider themselves correctors of a flaw in human nature. Malthus's followers proceed from different principles, but they also think that people reproduce too rapidly and want to correct this; they need only join the Skoptsy.[74]

Some mistakenly considered Darwinism to be a theory of disharmony in nature. The essence of Darwinism, however, was that all organisms arose from a few original forms as a result of "the influence, or more correctly, the pressure, of surrounding conditions."[75] The different conditions prevailing at different times and places had resulted in a branching tree of evolution. If conditions did not change, then neither could organisms; if conditions changed very rapidly, organisms perished. Properly understood, therefore, Darwinism provided a sharp refutation to the theorists of disharmony.[76]

Here, clearly, Beketov was invoking Darwin while describing his own theory of evolution. Indeed, in his attempt to reconcile Darwin's views with his own concept of harmony in nature, Beketov summarized the "entire essence of [Darwin's] theory" without mentioning either the struggle for existence or natural selection.[77]

Harmony, Not Struggle, 1882-1896

By the early 1880s Beketov had become convinced that Darwin's Malthusianism—his reliance on overpopulation and incessant intraspecific conflict—rendered his concept of the struggle for existence fundamentally incompatible with a belief in natural harmony. This conclusion and the critique of the selection theory that flowed from it were evident in his *Textbook of Botany* (1880-1883) and his essay "Darwinism from the Perspective of the General Physical Sciences" (1882).

Beketov introduced his *Textbook of Botany* with a philosophical discussion of the goals and limitations of biology. Building on the physical sciences, biology sought an "explanation of the mechanical activity that results in the phenomena peculiar to organic forms."[78] These explanations were limited to proximate causes and could not be extended to the spiritual realm, the essence of which would remain "forever unknown."[79] Modern evolutionary theory was therefore fully compatible with religious beliefs. While naturalists had formerly attributed the appearance of life forms to an endless series of miracles, "according to evolutionary doctrine the miracle consists in this: God created matter capable of altering itself without limit, and containing within it *in potentia* all the forces and forms corresponding to the immutable laws that He Himself established."[80]

Modern evolutionary theory rested primarily on Lamarck's insight that organisms adapted to the conditions surrounding them and on the inheritance of acquired characteristics.[81] Darwin had added the principle of natural selection through intraspecific competition (which Beketov termed "life competition"). Although of secondary importance, Darwin's contribution had helped "provide the theory of evolution with a solid basis."[82]

Beketov had come to identify "life competition" with disharmony and to minimize its evolutionary significance. There was, he observed, little evidence that "life competition" played an important part in the life and death of organisms:

> It is not difficult to accumulate thousands and tens of thousands of cases demonstrating that the number of offspring produced by plants and animals is immeasurably greater than the number that remain alive, but the essence of the question lies elsewhere. It is necessary to prove that these deaths are determined specifically by life competition, since the doctrine of natural selection is based on the proposition that new forms are developed through its activity.[83]

Where it did exist, "life competition" was not perpetual but eventually ended in balance. Drawing upon his colleague K. F. Kessler's article "On the Law of Mutual Aid" (1880), Beketov identified this ultimate balance with cooperation. "In this manner a phenomenon that begins with competition, with struggle, ends in mutual support and balance. So, if one wishes to use the expression 'struggle' it is more correct to speak of the struggle for balance than of the struggle for existence."[84] He provided an example from botany:

> According to the Darwinists, Malthus's law applies not only to humanity but to all organic nature as well. But natural selection and life competition lead to the weakening of imbalance and finally to the establishment of complete balance in the environment. A young wood, sown in a given field, presents an example of life competition ending in complete balance. In the beginning the saplings are set one around the other and seize nutrients from one another. Many of them perish, but finally, attaining maturity, they not only cease to interfere with one another but help each other to more effectively resist winds, preserve moisture and prevent settlement by foreign plants.[85]

However one-sided the Darwinist view of struggle and intraspecific relations, its moral implications were generally salutary. For one thing, it provided a decisive refutation of vulgar materialism: Darwin had demonstrated that human evolution led to the development of a "psychic type" who derived satisfaction from cultural, scientific, and religious pursuits that did not bring him into conflict with others.[86]

By the 1890s, however, Beketov had concluded that Darwin's errors regarding the struggle for existence rendered his theory both logically flawed and morally pernicious.[87] In "Morality and Natural Science" (1891) and *The Geography of Plants* (1896), he presented an elaborate critique of Darwin's metaphor—unpacking it, criticizing its Malthusian components, and juxtaposing it with his own vision of natural harmony.[88]

Darwin had appropriated the term struggle for existence from Malthus, Beketov explained, but used it much less precisely than had the political economist. In the third chapter of *Origin of Species*, "two different circumstances are confused, specifically: *the relationships of organisms to external physical agents*, which one can term the actual struggle for existence, and *the antagonism between organisms themselves*—life competition, properly speaking."[89]

The "actual struggle for existence" differed fundamentally from "life competition." The former transpired whenever an organism existed in an environment. The latter, however, depended on two additional elements: reproduction in an unlimited progression and limited space. Beketov drew on an analogy from mechanics: air always expanded on contact with heat, but expanding air did not always result in steam power. Steam power was created only within a structure that confined and directed the expanding gas. Similarly, the "actual struggle for existence" required only conflict between an organism and an environment; life competition required two additional natural mechanisms (unlimited reproduction and limited space) that confined and directed that conflict. The struggle for existence—conflict between an organism and physical conditions—was, then, "something physically unavoidable, since life itself is nothing other than reactions to external forces." Life competition—conflict among organisms—on the other hand, "does not constitute a physical necessity." It was, therefore, a mere "phenomenon" rather than a law of nature.[90]

The term "life competition" could be unpacked still further, revealing other ambiguities:

> The concept of life competition cannot include the struggle between heterogeneous organisms, since these relate to one another as combatants [*boriashchiesia*] and not as competitors [*sostiazaiushchiesia*]: the sheep and the wolf that attacks it do not compete, but rather struggle; man struggles with the tapeworms that inhabit his intestines, he does not compete with them.[91]

This distinction enabled one to avoid "the conceptual confusion so noticeable among Darwin and his followers" every time they attempted to deal with actual events in the plant and animal worlds:

> So, if two hunting dogs are engaged in battle with a wolf, we must ask ourselves, from the Darwinists' perspective, who is struggling here? The dogs with the wolf, or the dogs among themselves; since the dog who is stronger and more adroit will remain among the living, having finally subdued the wolf, but the weaker and less adroit dog will perish. This means [for the Darwinists] that the strong dog overcame the weak one! From our own point of view we will say without vacillation that both dogs struggle with the wolf, and are themselves in competition.[92]

For Beketov, then, the different processes conflated within Darwin's metaphor were properly classified as follows: (1) the organism's conflict with physical conditions (the "struggle for existence"), (2) indirect competition within a species ("life competition"), and (3) direct combat between members of different species.

Note Beketov's assumption that conflict within a species was always indirect while that between different species was always direct. This, I think, reflected his belief in the overriding importance of membership in a group (here, species). Relations among group-members were fundamentally different from those between nonmembers. He even suggested that the relationship between members of two different species had more in common with the conflict between an organism and the physical environment than it did with the relationship between two members of the same species.[93]

At the heart of Beketov's analysis of life competition lay his anti-Malthusianism. In his unpublished notebooks on "Morality and Natural Science," he observed that

> Malthus's stupidity has yielded dangerous fruits. Malthus concludes that widespread hunger, deadly epidemics and destructive wars will save humanity from perishing, postponing this until distant times. The poor, according to Malthus's counsel, should not even reproduce, or should do so with extreme caution. All these sordid principles, unhappily, gain further support by the incorrect framing of

the question about the struggle for existence.... If beasts were able to reason perhaps Malthus might be a good advisor to them, although one could doubt even this, but for beings endowed with the ability to reason such advice is extremely repulsive.[94]

So Darwin's incorrect approach to the struggle for existence lent support to Malthus's reprehensible ideas. Beketov made this point more politely in "Morality and Natural Science" (1891), where he related the Darwinist emphasis on life competition to the "blind, I will even say morally repugnant, conclusions which Malthus drew from his false law, and which many even today take for honest coin."[95]

Like most Russian commentators, Beketov distinguished between Darwin, who had never explicitly endorsed Malthusian social prescriptions, and "Darwinists," who did so enthusiastically. Like the Bible, the *Origin* was often invoked to justify crimes that would have appalled its author. Yet Darwin's Malthusian formulations had prepared the ground for such misinterpretations, for arguments that "science and the great Darwin" sanctioned "fierce egoism" and discouraged reproduction among the poor.[96]

Beketov's case against the evolutionary significance of life competition did not rely, however, on ethical pronouncements. He repeated and elaborated his argument of 1882 that Malthus's law was but a statistical abstraction of questionable relevance to actual events in nature.[97] Darwin himself had acknowledged that billions of organisms perished from frost, drought, and flooding, and that in extreme climates organisms struggled only against physical conditions. It was therefore obvious that "one cannot attribute every case of the annihilation of organisms to life competition, as this occurs only under conditions of limited space and surplus population." Where these conditions were not evident, "the so-called Malthusian law is not applicable."[98] Life competition, then, was not a law of nature but an "occasional phenomenon." It transpired only under specific conditions whose prevalence was often assumed but never demonstrated.[99]

Having redefined the struggle for existence to exclude the component most important to Darwin, and having relegated that component to the status of an "occasional phenomenon," Beketov devalued its evolutionary role still further. Even when life competition did occur, he asserted, it rarely resulted in the extinction of one species or the emergence of a new one. More frequently, it represented but a temporary perturbation, like the motion of a balance, and culminated in a new state of equilibrium.[100]

Furthermore, relations within a species were conditioned primarily by the common struggle against physical conditions. This often mitigated competition and led to mutual aid:

> Among social organisms antagonisms are weakened as a consequence of the necessity to struggle with surrounding conditions. From a general physical point of view this is the joint action of homogeneous forces having an identical direction and applied to one and the same point. Two horses jointly dragging the same burden are not at all in competition: obviously, they are aiding one another....
>
> It is sufficient to direct one's attention to our broad woods and meadows, consisting of one and the same species for time immemorial, to appreciate the mutual aid that plants provide to one another.[101]

Beketov applauded Darwin for mentioning cooperation in his *Descent of Man* but lamented Darwin's failure to appreciate the extent to which mutual aid vitiated life competition and became a source of harmony in nature.

This analysis of the struggle for existence left little of the selection theory. In a short autobiography published in 1891, Beketov recalled that after some initial enthusiasm for Darwin's ideas he had "become increasingly convinced" that the views expressed in his own "Harmony in Nature" were closer to the truth.[102] Most scientists, he claimed, also rejected Darwin's contention that evolution resulted "almost exclusively from natural selection":

> One can say that the greater part of Darwinian propositions have either been cast in a new light or even changed. The major principle of the theory [evolutionism], without doubt, remains firm, but is presently understood in a manner closer to the views of Lamarck and Saint Hilaire.[103]

Beketov's description of this modern consensus echoed the views expressed in his "Harmony in Nature" over thirty years previously:

> At various points on the earth general physical conditions were grouped and are grouped entirely differently, both quantitatively and qualitatively; at one and the same point, but at various times of the year, in various months, weeks, days, hours, the combinations of these conditions change. The activity of all these various combinations generated the enormous number of plant and animal species that now populates our earth. Each plant is not only the result of the action of a certain qualitative and quantitative relationship, a complex of external conditions, but is the direct morphological expression of the activity of a certain number and quality of external forces.[104]

The key task facing evolutionary theorists, then, was to uncover the mechanisms by which the environment elicited variations in the organism. For this information, scientists looked to the promising new field of experimental morphology.[105]

Celestial Harmony and Non-Malthusian Biology

As he neared the end of his career in the 1890s, Beketov could take great satisfaction in his accomplishments. Russian botany had come into its own, and his patrimony, as teacher, organizer, and scholar, was well-recognized. Twice in this decade he reflected publicly on the significance of his scientific work, and each time he emphasized his resolution of the problem of harmony and struggle in nature.

In the didactic short story "Doctor Froman" (1892), the sixty-seven-year-old Beketov's title character embodies the ideal relationship between scientific knowledge, philosophical wisdom, and Christian charity. Doctor Froman is an elderly ascetic who devotes almost all his waking hours to helping the poor. A modicum of self-indulgence is unavoidable among humans, and Froman's takes the form of a passion for science, especially astronomy.

One day a young science student offers to help Froman with his philanthropic pursuits. Tortured by a "dark, pathetic worldview" and impressed by Froman's peaceful demeanor, the student one day confesses his world-weariness. The young man is a zoologist, a narrow specialist who knows a great deal about annelids but has studied little else. Nature seems to him a depressing arena of "unending struggle and mutual extermination, with dire poverty almost everywhere." He sees only "this iron law of necessity: every being seeks to preserve its own existence, and to even a much greater degree, that of its species, and to do so even at the greatest cost in suffering and death."[106]

Froman leads the zoologist to his telescope and bades him look at the heavens. A specialty is a fine thing, Froman assures him, and is even necessary for the progress of science, but one can not understand the natural order by examining a drop of water or a cluster of crawling beasts. Astronomy reveals the big picture: the "universal balance which, despite perpetual motion, governs the heavens." Temporary perturbations in the skies are perceptible only to a trained observer and do not disturb their basic harmony. "The very struggle, the very discord among living beings, among humans, are also but temporary phenomena leading to harmonious balance. The Creator did not establish immutable laws in order that chaotic phenomena would predominate over harmonious ones."[107] Geology reveals that the cataclysmic volcanic activity of previous epochs have gradually yielded to the present, relatively stable "balance between internal forces and the crust of the earth." Inequality, poverty, and struggle among humans are also but temporary violations of universal harmony. Gifted with reason, man, through science, would eventually uncover the laws of harmony, attain Christian morality, and eliminate the privation and inequities that led to discord.[108]

Like his fictitious Dr. Froman, Beketov considered harmony, not struggle, the essential fact of nature. He defended this view, first by incorporating and finally by rejecting Darwin's vision of incessant conflict. The importance that Beketov attributed to this conclusion was evident in his autobiography, which ended with the following summary of his contribution to evolutionary theory:

> He recognized that the very conception of the struggle for existence was incorrect, demonstrating that *life competition*, the part of this phenomenon to which Darwin attributes the greatest significance, should be separated from the general concept of the struggle for existence—and that it is not a struggle for existence but rather a struggle for balance, or, more correctly, competition for balance. In this manner the very expression "struggle" should be discarded and Malthusianism loses its significance.[109]

Not all of Beketov's students, of course, accepted this part of his intellectual legacy. Timiriazev, for example, became Russia's most vociferous champion of Darwinian orthodoxy. For him Beketov was a great teacher and organizer of science, and the author of an early evolutionary tract that had "prepared the soil for Darwinism."[110] As for Beketov's analysis of the struggle for existence—this Timiriazev passed over in silence. Not so V. L. Komarov. Eulogizing his mentor in terms that Beketov would no doubt have appreciated, Komarov praised his ability "to present all the better aspects of Darwin's theory without descending into panegyrics, and to identify all its weaknesses and confusions without condemnation."[111]

CHAPTER 4

Korzhinskii, the Steppe, and the Theory of Heterogenesis

From a blind proponent of the struggle for existence [Korzhinskii] later suddenly transformed himself into a no less blind opponent of it.[1] (K. A. Timiriazev, 1905)

If the reader is familiar with Sergei Ivanovich Korzhinskii (1861–1900), it is probably as the Russian Hugo de Vries, as the man whose "Heterogenesis and Evolution" (1899) appeared two years before the Dutch biologist's *Die Mutationstheorie.* Indeed, Korzhinskii's contribution to evolutionary theory attracted more comment at Cambridge University's Darwin celebration in 1909 than did the work of any other Russian scientist.[2]

This no doubt galled Timiriazev, for Korzhinskii may well have been his least favorite Russian naturalist. For Darwin's leading Russian champion, Korzhinskii's theory reflected his transformation from a "blind proponent" of Darwin's concept of the struggle for existence to "a no less blind opponent of it," a metamorphosis attributable to unbridled careerism and political opportunism.

Timiriazev's accusation was almost certainly unfounded, but it did identify a single theme that united Korzhinskii's early research on botanical geography with his later work on evolutionary theory: an analysis of the nature and significance of the struggle for existence. As for Korzhinskii's transformation from "proponent" to "opponent" of Darwin's concept, we shall see that this reflected an underlying philosophical consistency: his conviction that organic phenomena were not the simple products of external influences. When addressing the set of issues involved in botanical geography, Korzhinskii found this philosophical position highly compatible with Darwin's struggle for existence, which he placed at the center of his own controversial analysis of the division between forest and steppe. When addressing the very different set of issues involved in evolutionary theory, however, he found that same Darwinian concept to be philosophically unpalatable, and emphasized one great advantage of his own brand of transformism: its denial of any creative evolutionary role whatever to the struggle for existence.

Enfant Terrible of Russian Botany

In Gogol's "Taras Bulba" the victorious Polish army tortured and executed a Cossack prince near the city of Dubno. The Korzhinskii family claimed its descent from that prince's surviving brother. By the time of Sergei Ivanovich's birth, in Astrakhan in 1861, this family of landless nobles had produced a number of architects, physicians, and other professionals.[3]

Little information is available about Korzhinskii's early life. At age thirteen he enrolled in the Astrakhan Classical Gymnasium, where he distinguished himself as a student, became particularly interested in botany, and participated in excursions along the Volga delta. In 1881 he entered Kazan University, where he studied botany with N. F. Levakovskii and worked closely with P. N. Krylov of the university's botanical garden.

During his student years Korzhinskii embarked on numerous expeditions throughout the Kazan area, especially along the northern boundary separating the forests from the rolling grasslands (steppes) of eastern Russia. Russian naturalists had long sought to account for the sudden transition from forest to steppe, and Krylov encouraged Korzhinskii's interest in this subject. Krylov left for Tomsk University in 1885, however, and Korzhinskii's adviser, Levakovskii, urged him to tackle a less controversial subject. Levakovskii supervised Korzhinskii's master's essay on the biology and geographical distribution of the rootless predatory plant *Aldrovanda vesiculosa* (1887) and urged him to expand it into a workmanlike and uncontroversial doctoral dissertation.

Rejecting this counsel, Korzhinskii submitted a doctoral thesis entitled "The Northern Boundary of the Black-Earth Steppe Region of the Eastern Zone of European Russia, with Respect to Its Soil and Botanical Geography" (1888, 1891). By contending that this boundary was not determined by soil and climatic conditions, but rather by the struggle for existence between steppe and forest formations, he dissented radically from the prevailing consensus. The twenty-seven-year-old Korzhinskii accused the authors of that consensus—the leading lights of Russian botany—of praising Darwin effusively while ignoring the essence of Darwin's concept of the struggle for existence.[4]

His dissertation defense was successful but unpleasant. The faculty committee was shocked both by his presumption in tackling such an important question and by his novel resolution of it. Levakovskii sat silently throughout the entire ordeal.[5] Korzhinskii had intended to dedicate his thesis to Krylov, but the faculty's reaction persuaded him to omit what seemed a dubious honor.

Disheartened by life in Kazan, Korzhinskii leapt at the opportunity to join Krylov in Siberia at the newly-created Tomsk University. It had, he confided in a letter to Krylov, become "difficult to breathe" in Russia. Intellectual life had withered under the restrictive University Statute of 1884. "The servility of our universities, the weeding out of everything dear and good in the university structure, the demoralization of science and teaching—these create a situation from which one turns away in horror. I am prepared to run away even to Sakhalin." The political and economic situation in Russia was "hopeless," and the status of science was intolerably low. "Conservatives consider it dangerous and conducive to free-thinking, and liberal youths think it unnecessary and useless; they equate scientific studies with careerism."[6]

Siberia offered a more promising setting for cultural and scientific pursuits. Perhaps Siberian youths would prove less subject to political extremism than their counterparts in Kazan, and so more likely to recognize science as "the single path of development for mankind." "We will be *Kulterträger* and if our work bears fruit we will enjoy life's greatest pleasure—a sense of productivity and usefulness."[7] Tomsk was also an ideal location for a botanical geographer. "Where else can one find such broad expanses for phytogeography as in Siberia," he enthused. "We will study primitive nature untouched and unmaimed by man (will it not be too crowded with the two of us?)."[8]

Upon his arrival in Tomsk in the fall of 1888, Korzhinskii delivered a controversial address, entitled "What Is Life?" (1888), to the university's commencement ceremony. This attempt to develop a position that respected the distinctive qualities of organic phenomena without reverting to the notion of "vital force" was widely perceived as vitalist.[9]

A popular lecturer and prolific scholar, Korzhinskii taught botanical geography and became increasingly interested in its implications for evolutionary theory. During his four years in Siberia he published thirteen scientific works, the most substantial of which were the second section of his dissertation (1891) and *The Flora of Eastern European Russia* (1892). His numerous expeditions included one, sponsored by the Eastern Siberian section of the Imperial Russian Geographical Society in 1891, to determine the feasibility of establishing Russian settlements in the Amur region.

His political conservatism and scientific achievements combined to bring him professional success in the 1890s. In 1892, at age thirty-one, Korzhinskii was appointed main botanist of the Imperial Botanical Garden in St. Petersburg. One year later he became director of the Imperial Botanical Museum and an adjunct member of the Academy of Sciences. He was elevated to full membership in 1898.

During his brief tenure as director of the Imperial Botanical Museum, Korzhinskii began to reestablish it as an important research center. He convinced the authorities to fund a massive inventory of Russian flora, to which he himself contributed over ten thousand specimens. At his initiative the museum also founded its own journal organ, *Travaux de Musée Botanique de l'Académie des Sciences de St. Péterbourg*, which commenced publication in 1902.[10]

In the middle and late 1890s Korzhinskii traveled throughout eastern and southern Russia, to Turkestan and the Caspian region. In 1898 he again turned his talents to practical use, accepting a government request to investigate means of introducing new types of grapes into the Crimea. The results were published in *Ampelography of Crimea* (1904).

Korzhinskii is best known to Western scholars as the author of the theory of heterogenesis. Drawing on his field experience and the observations of foreign botanists, he advanced this alternative to Darwin's theory in a speech to the Academy of Sciences in 1899. The following year he published the first, largely descriptive section of a comprehensive statement on his mutation theory; his work on the second, theoretical section was abruptly terminated by his death from a kidney infection in 1900.

After his death Korzhinskii was praised highly by such botanists as A. Ia. Gordiagin and A. S. Famintsyn, and by the biologist Iu. A. Filipchenko, who considered him one of only four nineteenth-century Russian scientists to make an important contribution to evolutionary biology.[11] Yet Korzhinskii was alternately ignored and vilified by Russian historians of science in subsequent decades. For one thing, Timiriazev had dismissed him as a dishonest and servile reactionary metaphysician, and Timiriazev's opinions carried great weight in the years after 1917. For another, Korzhinskii's political conservatism, his criticism of Darwin, and his rejection of the philosophical materialism of his day hardly made him an attractive figure for historians working amid highly politicized Soviet biology in the 1930s, 1940s, and 1950s. In 1961, during the Khrushchev years, G. D. Berdyshev and V. N. Siplivinskii produced a short volume that remains the only biography of Korzhinskii. While criticizing his philosophy and his theory of heterogenesis, the authors praised this "first Siberian botanist" for his contributions to botanical geography and to the Imperial Botanical Garden. And while avoiding discussion of Korzhinskii's political views, they exonerated him of Timiriazev's charge that he had changed his attitude toward Darwin's theory in return for a state subsidy.[12]

Why Is the Steppe Treeless?

Russian naturalists often observed that their country, by virtue of its dramatically contrasting geographical and botanical regions and its vast expanses unaltered by human exertions, provided an ideal setting for studying the relationship between life forms and their environment.[13] This relationship lay at the center of a question that preoccupied Russia's botanical geographers in the nineteenth century: Why did the vegetation of Russia's steppe and forest regions differ so radically?

Aside from its theoretical import, this question had important implications for agriculture, industry, and settlement policies. So the St. Petersburg Society of Naturalists, the Imperial Free Economic Society, and the Ministry of State Domains subsidized numerous expeditions throughout the steppe and forest regions, and especially along their common boundaries. These expeditions fueled a rich discussion about the geographical distribution of plants—a discussion in which differing conceptions of the struggle for existence played an important and explicit role after Korzhinskii challenged the dominant view in his publications of 1888 and 1891.[14]

Few Russian naturalists would have quarreled with the description of the forest and steppe regions that was offered to readers of the *Encyclopaedia Britannica* in 1886 by the expatriate Russian geographer and geologist P. A. Kropotkin. "The Forest Region of the Russian botanists occupies the greater part of the country, from the Arctic tundras to the Steppes," Kropotkin observed, "and it maintains over this immense surface a remarkable uniformity of character."[15] Beketov had properly divided this vast area into the forest region proper and the antesteppe. The forest region proper was divided into two parts, the coniferous and oak forest zones. The coniferous region lay to the north and featured large, impenetrably dense forests, marshes, and thickets, interrupted frequently by lakes and swampy meadows, with occasional clear and dry spaces. To the south lay the oak region, which covered all of central Russia and had a totally different character. Its undulatory surface lacked the marshy meadowlands of the north, and the dry soil of its forests hosted pine as well as birch, oak, and other deciduous trees.

The antesteppe constituted an intermediate zone, south of which lay the steppe proper.[16] This was a region of

> very fertile elevated plains, slightly undulated, and intersected by numerous ravines which are dry in summer.... Not a tree is to be seen, the few woods and thickets being hidden in the depressions and deep valleys of the rivers. On the thick sheet of black earth by which the Steppe is covered a luxuriant vegetation develops in spring; after the old grass has been burned a bright green covers immense stretches, but this rapidly disappears under the burning rays of the sun and the hot easterly winds. The colouring of the Steppe changes as if by magic, and only the silvery plumes of the feathergrass (*Stipapennata*) wave under the wind, giving the Steppe the aspect of a bright yellow sea. For days together the traveller sees no other vegetation; even this, however, disappears as he nears the regions recently left dry from the Caspian, where salted clays ... take the place of the black earth.[17]

Strictly speaking, the steppe was not totally devoid of trees. Clusters of wild fruits and other deep-rooted shrubs grew in the depressions of the land and along the slopes of ravines, "giving the Steppe that charm which manifests itself in the popular poetry." Rich vegetation grew in the marshy bottoms of ravines and in the valleys of streams and rivers, and the deltas of the Black Sea rivers sheltered a forest fauna. Here appeared an occasional oak forest, as well as a number of species unknown in the forest region, including the maple, hornbeam, and poplar.[18]

Why were there so few trees in the steppe?[19] The prevailing view among soil scientists and botanists was that physical conditions there were simply unsuitable for forests. Within this broad consensus there were considerable differences in emphasis. K. E. von Baer (1856) attributed the relative treelessness of the steppe to the inadequacy of rainfall, Middendorf (1864) to the terrible winds and dry air, P. A. Kostychev (1886) to the inability of the coarse-grained, black-earth soil to retain moisture, and A. N. Krasnov (1891) to the effects of receding moisture after the Ice Age.[20]

Beketov was both an architect and a typical representative of this consensus. His opinion, first expressed in 1877, remained essentially consistent throughout his career. He explained to the St. Petersburg Society of Naturalists:

> The treelessness of the southern Russian steppe ... arose as a consequence of the following: the land, having emerged from the ocean, was saline and consequently not suitable for tree growth. When, as a consequence of the gradual lexiviation of the salt, the saline grasses covered the land, this attracted herds of grazing animals, obstructing the development of tree growth, which would have been opposed anyway by the major factor in the distribution of plants, specifically the climate of the Russian steppe, which is far from propitious for forests.[21]

In expeditions sponsored by the Imperial Free Economic Society, the Imperial Botanical Garden, and the St. Petersburg Society of Naturalists, a number of Beketov's students, including G. I. Tanfil'ev and V. Ageenko, elaborated this hegemonic view that physical conditions in the steppe prevented the growth of forests.[22]

Levakovskii's Experiments

At Kazan University Korzhinskii's future advisor, N. F. Levakovskii, conceived a somewhat different approach to botanical geography after reading Darwin's *Origin of Species*. Levakovskii fastened on Darwin's concept of a multisided struggle for existence and decided that, once experimentally clarified, this concept could prove useful to plant geographers. In 1869 the Kazan Society of Naturalists accepted Levakovskii's proposal to investigate the dynamics of the struggle for existence among wild plants. A tract of land behind the university was chosen, and his experiments continued for about three years.[23]

"It is easy to see the extraordinary complexity of the question of the so-called struggle for existence," Levakovskii observed, and "an attempt to discover the several causes facilitating the supplanting of some plants by others is possible and not at all lacking in interest."[24] His experiments involved two sets of trials. In the first he tested the capacity of different seeds to absorb moisture at varying temperatures and in different types of soil. In the second he planted germinating seeds of different species together in a series of boxes. These boxes contained a variety of different soils and were exposed to varying amounts of sunlight and moisture. Thus, Levakovskii could observe the struggle for existence under a variety of environmental conditions.

He began his first report, "On the Supplanting of Some Plants by Others" (1871), with a general discussion of the struggle for existence that followed Darwin's argument closely. Levakovskii agreed that the struggle for existence resulted in part from population pressures, that it was sharpest among members of the same species, and that its intensity was regulated by a number of factors. He categorized the many factors mentioned by Darwin and added one more: the more nutriment contained in a seed, the more intense the struggle, since this led to more plants with greater vigor. His long discussion of the role of climatic factors was drawn almost verbatim from the *Origin*.[25] But there were two important differences between Darwin's and Levakovskii's theoretical comments: Levakovskii did not mention Malthus, and he emphasized the need to assess experimentally the relative importance of different types of conflict.[26]

Levakovskii then related the results of experiments that, he contended, suggested that differences among organisms played an important part in their relative ability to survive under different conditions. For example, germinating seeds varied greatly in the quantity of water that they absorbed and the rapidity with which they did so. His experiments had revealed marked differences, not only between the seeds of different species, but also among those of individuals of the same species. Thus, he commented, there was every reason to believe that the distribution of plants was not determined by physical conditions alone. Plants could, of course, survive only within a limited range of physical conditions. There must, for example, be some correspondence between the soil's ability to retain moisture and the type of plants growing on it. But it was equally clear that where these minimal conditions existed, the survival of some forms and the death of others must in large part result from differences among the plants themselves.[27]

This became the central contention in his second article, published in 1873. Having already established the range of physical conditions within which each plant species could survive, he now attempted to demonstrate that, within this range, the struggle for existence between plants determined the population of a given locale.[28]

For example, he described one trial in which he planted pine and birch seeds together in several boxes and maintained soil and moisture conditions suitable to both:

> Some of the boxes . . . were placed under direct light, others in the shade of some large trees. In both trials the pines germinated first, and the birch only afterwards. But the early appearance of the pines had much different consequences in the sunny and shady places: in the former, the young pines rapidly perished, surrendering their place to those birch that remained; and these latter, consequently, became predominate. Things proceeded otherwise in the shady trials: the pines, having germinated first, developed and waged a stubborn struggle against the young birch shoots, as a consequence of which many of the latter were compelled to surrender their place.[29]

Beketov and others might well have interpreted this as proof that the extent of sunshine determined whether the pines or the birches would prevail. For Levakovskii, however, this trial demonstrated that in physical conditions suitable for either species, the outcome depended on "determinate relations based on the ability of each species to conduct a struggle with other species."[30]

We should note that although Levakovskii accepted Darwin's emphasis on intraspecific conflict, and had confirmed experimentally the existence of significant differences within a species, he was here emphasizing interspecific struggle. The reason for this is clear: he hoped to illuminate the geographical distribution, not the evolution, of plant forms. The key question, therefore, was to what extent the distribution of plants is determined by a species' relationship to physical conditions and to what extent by its relationship to other species.

Levakovskii suggested that his results had great significance for explanations of temporal changes in regional flora, and particularly for an understanding of the relationship between forest and steppe. He recalled Middendorf's observation that in Siberia past generations of trees were buried under contemporary forests composed of completely different species, and also alluded to the well-known fact that when steppe land in southern Russia was left fallow it was occupied by a procession of different plants. These observations accorded with an approach to plant geography that emphasized the struggle for existence among plants rather than the direct struggle between plants and physical circumstances.[31]

Although Levakovskii's reports addressed a highly controversial subject, they elicited no reaction from botanical geographers. These naturalists were, after all, accustomed to wide-ranging field investigations, and Levakovskii offered only a narrow experimental study. In his doctoral dissertation Korzhinskii would develop Levakovskii's central contention—that the struggle for existence among organisms was central to botanical geography—and he would do so by addressing the controversial forest-steppe question in a language familiar to botanical geographers.

Korzhinskii on Steppe and Forest

The theoretical point of Korzhinskii's dissertation (1888, 1891) was advanced forcefully in his polemical conclusion:

> To ignore the life qualities of competing forms and, while giving lip service to the struggle for existence, to actually attribute all phenomena to climatic or soil conditions is, in my opinion, a great logical error. Much has been written about the struggle for existence. Everyone speaks with more or less enthusiasm about its singular importance to the origin, extinction and distribution of organisms upon the earth. Nevertheless, at least with respect to botanical geography, it remains a mere abstraction, since in the investigation of specific questions this struggle for existence usually remains peripheral, overshadowed by other factors.[32]

For Korzhinskii the great merit of Darwin's approach to the struggle for existence was his emphasis on organism-organism relations, his recognition that vegetation was not the passive object of environmental circumstances but was, rather, "something distinctive [and] independent." Plants actively transformed their physical environment, and the relations among them often led to changes in regional flora even when physical conditions remained constant. So the botanical geographer who truly grasped Darwin's concept of the struggle for existence would no longer insist on explaining the distribution of plants by physical conditions alone. He would, rather, emphasize the "gradually developing social relations among various [plant] forms."[33]

Scholarship on the steppe-forest question, Korzhinskii argued, consistently and dogmatically failed to take these relations into account. For example, at the urging of Beketov and soil scientist V. V. Dokuchaev (1846–1903), the St. Petersburg Society of Naturalists had equipped an expedition to study the relationship of the flora and soil of the Nizhegorod region. In their reports to the society in 1884–1885, all three of the expedition's naturalists had concluded that the plants flourishing in Nizhegorod's black-soil did not require black-soil at all but could grow on other soils as well. Committed to the view that plant life was the simple product of physical circumstances, however, the three had blithely assumed that the presence of these plants in black-soil regions and their absence in others must be due to climatic conditions.[34]

Korzhinskii insisted that the plant cover of a locale resulted also from the relations among plants, and that this aspect of the struggle for existence provided the key to vexing issues in botanical geography:

> There arises a competition between species of plants, there begins a struggle for existence, in which the weak individuals, and many species, perish. As a result of many centuries of this struggle for existence there develops in each land, from among those species which are most lively and best adapted to the given climatic and topographic conditions, special combinations of forms, constituting a so-called *plant formation*. These formations are essentially stable plant communities; they

occupy a territory and are subject to changes and the influx of new forms only when extraordinary circumstances destroy the normal conditions of plant life.[35]

Two fundamentally different plant communities had arisen in the Kazan region: the forest and steppe formations. Forest formations originated with the growth of trees. These altered the temperature and moistness of the air, and transformed the character of the soil by occupying certain strata with their roots and leaving others free. These conditions, in turn, favored some types of vegetation over others, and so an integrated, collective "organism" arose. The vegetation of steppe formations was similarly conditioned by the absence of trees.[36] Those who attributed the flora of the steppe regions simply to a combination of climatic and topographical conditions were, then, profoundly mistaken. There was no evidence that the plants of Russia's steppe region could not also thrive under the physical conditions of other areas. Rather, "the basic law conditioning the distribution [of the steppe] is the antagonism between steppe plants and the more powerful, more complete forest formations."[37]

What made the forest formation "more powerful"? Its component parts were more tightly interconnected, making them "a more complete, more individualized organism." In the steppe formation, on the other hand, "we don't observe . . . such a subordination of elements, since within it all members enjoy equal rights [*ravnopravnye*] and each member develops in identical conditions of light, heat and moisture."[38]

So while the potential geographical range of plants was established by physical conditions, their actual distribution reflected a struggle for existence among them. This struggle had a historical dimension—it was age-old and ongoing—and so the plant cover of a region often changed radically amid relatively stable physical conditions.

For Korzhinskii the current distribution of steppe and forest formations reflected one historical moment in the continuous struggle between them, a struggle in which the forests were gaining ground inexorably due to their more powerful organization. The steppe flora existed only where the forest had not yet penetrated or where physical conditions completely prohibited its growth.

> The distribution of forest and steppe formations does not depend directly on climate, nor on the topographical character of the locale, nor on the nature and characteristics of the substrata, but only on the conditions and course of the mutual struggle for existence.
>
> But it can be objected: the result of the struggle for existence must be conditioned by something. Oh, without question. The result . . . will depend always on two spheres of phenomena: first, on the relations of the competing forms with external conditions . . . and second, on their internal, specific life qualities, which give them greater or lesser chances in the struggle. Each organism presents certain needs to the environment . . . . And if the physicogeographical particularities of the locale do not provide these necessary conditions . . . then it stands to reason that the organism cannot exist. But physicogeographical conditions do not constitute the entire life conditions of the plant; there remains the entire world of the social relations with other organisms. It can easily occur that the climate and other physicogeographical conditions are entirely favorable to some form, but, nevertheless, it cannot exist in a given place because it is supplanted by more powerful compet-

itors. In another place, perhaps with incomparably less fortuitous climatic conditions, this form prospers, since a different selection of competitors gives it the possibility of seizing the advantage over them. Therefore, one cannot conclude from the absence of some species in a land that the climatic or soil conditions there are unsuitable for it.[39]

Russian botanical geography, then, required a new orientation, one based on Darwin's approach to the struggle for existence. The puzzles of the plant world would be resolved through a determined "investigation of the struggle for existence, of all the conditions and factors providing an advantage to one or another species."[40]

Korzhinskii illustrated his point in a chapter concerning the origin and fate of oak forests in central Russia.[41] In the summer of 1886 he had accompanied the forest inspector of Kazan province, K. M. Patkanov, on a tour of the area's oak forests. In an obvious dig at Beketov, Korzhinskii related Patkanov's conclusion that the depredations of rabbits, who gnawed at oak shoots, did not prevent these shoots from eventually growing into strong, healthy trees. It was therefore puzzling that a search of many hours failed to locate even a single young oak tree in the forest.

Korzhinskii offered the following explanation: It was well-known that oaks thrived on sunshine. As an oak forest grew, it gradually shaded the soil beneath it. This canopy thickened as the oaks matured, and the shade prevented a second, replacement generation from arising. Only if an older oak died could a young oak thrive in the resultant sunlight, and even then it was often stifled by rapidly growing weeds. More frequently, a "shade-loving" species thrived in the shadow of the oaks, struggled successfully against the weak efforts of a second oak generation, and dominated the forest as the first generation of oaks died. Their death, however, exposed the forest to sunshine, facilitating another exchange of species. This illustrated "the replacement of one species by others, without any changes in the soil, climate or other environmental conditions, as a result solely of the life qualities of the competing forms."[42] This admittedly a priori analysis was supported by many observations by foresters, who had informed Korzhinskii that oak forests often failed to renew themselves even under optimal physical conditions. Foresters had also described the difficulty of maintaining oak trees in a hybrid forest, where they were supplanted by trees that thrived on shade.

Again alluding to Beketov, Korzhinskii added that some saw these hybrid forests as

> a combination of forms which are in balance as a result of the climate or some other external conditions. Oh, no! We have before us only a transitional stage, one of the phases of a struggle, the result of which is not difficult to foresee. If the oak lives together with one of the more shade-loving species, then we are witnessing only one of the stages in the transformation of an oak forest to that of another species. And there can be no doubt that under normal conditions, that is, without human interference, this transformation will occur and the oak will be supplanted. It is, of course, quite another question whether this other species, having supplanted the oak, will in its turn surrender its place to other, more stable, lively forms. And perhaps an entire series of such replacements will occur continually, without any changes in climate or other physicogeographical conditions.[43]

Oak forests, he concluded, could arise in only one of two ways: they could grow in unoccupied places or they could replace trees that were even more "sun-loving" than themselves.

This explained the presence of scattered forests, sometimes containing oak trees, in Kazan's steppe regions. Korzhinskii dismissed as "a strained, not to say naive, interpretation" the dominant view that these reflected "certain combinations of external conditions."[44] They attested, rather, to the gradual afforestation of the steppe, to a historical process, dependent on the struggle for existence between plant species, that began with the growth of oaks on the steppe and continued through their replacement by a procession of other species.

Russia's plant geographers should read Darwin more closely, Korzhinskii concluded, for their errors illustrated the wisdom of Darwin's observation that "nothing is easier than to admit in words the truth of the universal struggle for life, or more difficult . . . than constantly to bear this conclusion in mind."[45]

"What Is Life?": Vitalism and the Steppe-Forest Question

In his doctoral dissertation Korzhinskii criticized commentators on the steppe-forest question for ignoring the significance of a plant's "internal, specific life qualities." He insisted that these qualities, which were independent of external, physical circumstances, ultimately determined the geographical distribution of flora. In the year that his dissertation was published, Korzhinskii upheld the same philosophical position in a speech about vitalism and mechanism delivered to the commencement ceremony at Tomsk University.

In "What Is Life?" Korzhinskii identified himself with the vitalist argument that organic activities could not be explained solely on the basis of physical and chemical laws. Plants and animals were active and purposeful. Their development and daily activities "do not depend entirely on the conditions surrounding them."[46] This testified to the presence in all organisms of "some deep, secret force . . . poured throughout the organic world, glimmering in every molecule of the plasm and blazing as a flame in human reason." This force constituted "the memory and activity of the plasm: it is life."[47]

Yet nothing was gained by a return to vitalism and its justly discredited notion of "life force." Korzhinskii proposed instead the term "life energy." This expression connoted the presence of a unique force among living bodies, a force "immeasurably more complex, more subtle" than those currently understood by physics and chemistry. Unlike "life force," however, "life energy" was neither spontaneous nor inexhaustible. It resulted from oxidation, had a material reality, and could not contravene the laws of the material world. Perhaps it would eventually be understood as the product of its constituent parts, perhaps it would not.

For Korzhinskii scientific progress was not a linear but a dialectical process. In some periods scientific theories leaned tendentiously toward mechanistic ideas and in others toward vitalist ones. The recent popularity of mechanism had served a useful historical role and had generated important knowledge about physical and chemical influences upon living bodies. Yet its usefulness was now exhausted. Biology now required a corrective adjustment toward the "share of truth" in vitalist theories.[48]

The essential point for Korzhinskii—and one that linked his concept of the struggle for existence to the views expressed in "What Is Life?"—was that the development and activities of organisms could not be understood simply as the result of external physical forces. His resolution of the steppe-forest question and his attack on prevailing views of the struggle for existence flowed from this general position that biological ideas had veered too sharply in the direction of mechanistic ideas.

The relationship between Korzhinskii's philosophical views and his resolution of the steppe-forest question was not lost on botanical geographers. Two aspects of the argument in his doctoral dissertation survived initial criticism to gain wide acceptance: (1) forests alter their physical environment, and (2) botanical geographers should concentrate on plant formations rather than individual plants.[49] His insistence that relations among plants constituted the central aspect of the struggle for existence, however, was widely criticized as vitalist.[50]

Among Korzhinskii's sharpest critics were Beketov and Dokuchaev, as well as the botanist G. I. Tanfil'ev (1857–1928), who had studied with them both. Dokuchaev responded almost immediately. An authority on the steppe, he had previously argued in *The Russian Black-Earth* (1883) that soil conditions exercised a determinate influence on regional vegetation. In 1888 he ridiculed Korzhinskii's contention that the forest had moved south and supplanted the steppes, that as "a result of the struggle for existence (?)" forests had changed the very character of the soil. "Where is the proof," he asked "for such a grandiose conclusion?"[51]

Tanfil'ev, a leading member of Dokuchaev's Soil Commission who had traveled widely throughout European Russia, addressed Korzhinskii's argument at length in his master's thesis, *The Boundaries of the Forests in Southern Russia* (1894).[52] Acknowledging Korzhinskii's profound impact on the forest-steppe controversy, his distinctive emphasis on relations among plants, and his praiseworthy demonstration that the forest was gradually supplanting the steppe, Tanfil'ev insisted, however, that Korzhinskii's approach to the struggle for existence was metaphysical.

What could it mean, he asked, to assert that the steppe was "supplanted by more powerful competitors," and that the forest was "more powerful and more complete" than the steppe? Only that it was better adapted to such environmental conditions as moisture, light, food, and parasites. The specific factors favoring one life form over another were always difficult to discern. Yet by invoking "internal, specific life qualities" instead of analyzing these external factors, Korzhinskii converted the struggle for existence into a purely abstract concept, one that actually obstructed scientific investigation.[53]

> Science is attempting to understand even such complex phenomena as the flights of birds, the migration of lemmings, and the extinction of animals in terms of specific external physical causes; even the more complex phenomena in the life of human societies are explained—sometimes entirely satisfactorily—by the influences of nature, that is, again, by such external causes. Nobody, of course, will reject the struggle for existence, but to recognize it in Korzhinskii's sense is to recognize, without any need to do so, that long outdated notion of life force, and to

explain a specific distribution of plant formations by an abstract struggle for existence is equivalent to refusing to explain it at all.[54]

Like Korzhinskii's comment about the debilitating "equality" within steppe formations, this passage reminds us that these discussions about nature involved perceptions about human society as well.

Tanfil'ev illustrated the dangers of Korzhinskii's position by refuting one of his most telling examples: the plant *Tussilago farfara*, which thrived in regions with widely varying soil and climatic conditions. Korzhinskii had made much of naturalists' failure to explain this plant's geographical distribution in terms of the determinant influence of soil and climate, and had offered a simple explanation based on the social relations between plants: Where stable plant communities exist, they resist the incursions of newcomers; *T. farfara* was never incorporated into plant communities and so existed only where such formations did not. For Korzhinskii, then, the distribution of *T. farfara* was explicable only in terms of the relations among organisms.[55]

Tanfil'ev responded that Korzhinskii's metaphysical "solution" merely codified ignorance. Although the geographical distribution of *T. farfara* was not dependent on the most evident characteristics of the soil, it turned out, on closer inspection, that it could grow only in land that was rich in carbon salts.[56] Thus, Korzhinskii's "geobotanical paradoxes" simply reflected the incompleteness of present knowledge, and his assertion that such paradoxes could be resolved only by invoking a highly abstract struggle for existence was a diversion from thorough scientific investigation. Tanfil'ev offered "a simpler and more natural explanation [of the gradual afforestation of the steppe] that does not require recognition of an active, distinctive and independent struggle for existence."[57] This rested on the changing lime content of black soil.[58]

In *The Geography of Plants* (1896) Beketov criticized Korzhinskii in much the same terms. Certainly, forests could present "a direct mechanical obstacle to the spread of meadow grasses" and could favor the growth of forest grasses. But this influenced only the topographical, not the geographical, distribution of plants. Climate was most important to the latter, and obviously a forest in moderate or cold climes would not obstruct grasses native to tropical ones.[59] He added:

> These facts are cited as cases of the struggle for existence. It is said, for example, that the forest struggles with the steppe grasses, etc. Some attempt to explain by this struggle certain facts in the topographical and even the geographical distribution of plants. Here, clearly, exact interpretation has been poorly served by the figurative expression "struggle." I have above analyzed sufficiently the influence of general physical conditions on the geography and topography of plants. In this case it is clear that the matter is resolved by the distribution of what I have termed the biological complex of plants. If a locale contains the optima of all the general physical factors of the forest or, more exactly, of one or the other trees, and . . . [a minimal amount] of these conditions regarding certain grasses, then that place will be covered by a forest. In the contrary case, it will be covered by grasses. It is easy to say that the forest has triumphed in the struggle for existence in one locale, and the grasses in another, but this is only a rhetorical flourish and not an explanation.[60]

For Beketov, Korzhinskii's analysis of the struggle for existence illustrated the dangers inherent to Darwin's metaphor.[61] Like Tanfil'ev he worried that such "rhetorical flourishes" might divert naturalists from the essential task of discovering how external conditions affected the physiological processes of plants, favoring some over others in a given environment.

Surveying the debate in 1894, plant geographer N. I. Kuznetsov concluded that most of Russia's botanical geographers were in basic agreement with Dokuchaev, Tanfil'ev, and Beketov. Although scholars now agreed with Korzhinskii that relations among plants deserved attention, the majority continued to see them as of distinctly secondary importance and to emphasize the determinate influence of physical factors.[62]

Timiriazev dismissed Korzhinskii's argument as a cover for idealism. He observed in 1895, "One can still find biologists who think that everything is explained by uttering the expression 'struggle for existence,' and who are prepared to regard with indignation and mockery . . . any application of physical methods of investigation to living beings."[63]

From Botanical Geography to Evolution: The Theory of Heterogenesis

Korzhinskii's contribution to botanical geography was distinguished by his creative and, for Russians, provocative use of Darwin's concept of the struggle for existence. His later contribution to evolutionary theory was distinguished by his refusal to concede any creative role whatever to that same struggle for existence.

This change of attitude seems less radical if we consider an important difference in the type of question posed by these two areas of interest. In botanical geography the issue for Korzhinskii was the relative importance of physical circumstances and interspecific struggle in determining the physical geography of existing plant species. Here he found in Darwin's struggle for existence an emphasis on relations among organisms that resonated with his own philosophical outlook—with his conviction that living beings were not simply the puppets of external physical forces. In evolutionary theory, however, the issue was whether struggle among organisms, especially among those of the same species, led to the emergence of new species, to a rich variety of living forms, and to progress. Here Korzhinskii found in Darwin's struggle for existence a morally pernicious doctrine that offended his deeply-felt beliefs by ascribing progress, not to the internal properties of organisms, but to harsh struggle and suffering.

He first addressed evolutionary issues in *The Flora of Eastern European Russia* (1892). In a long introduction he railed against the lifeless empiricism of botanists and urged them to address two general questions. First, "in what external conditions and in what social grouping" do specific plant forms dwell? Second, what are the origins and evolutionary paths of these forms? Vegetation, after all, was not "dead and lifeless," as it too often appeared in textbooks. It was "something alive, in constant motion, subject to continual, constant transformations, with a history, a past, and a future."[64]

Plants were both passive physical entities and active social beings. This perception lay at the heart of Korzhinskii's comments on the nature of species. He distinguished between a species' external physical characteristics, or *morfoma*, and its "true essence," or *biont*. The nature of the *biont* could not be discerned by examination of the *morfoma* only:

The *biont* of a species characterizes an entire series of other specific qualities, both sexual and social relations with other forms, the length of periods of development, certain reactions to climatic and soil conditions, etc. In order to understand a form one must not limit oneself to the *morfoma* but must also study all the external manifestations of its *biont*. In this manner its degree of individuation will be explained. The question "Does it constitute an individual species or subspecies or variation?" is, actually, a question of the degree of the individuation of its *biont*.[65]

So, as the *biont* expressed itself with increasing clarity over time, a race developed subspecies and, finally, new species.

The same concern with plants as the expression of both external and internal forces was evident in Korzhinskii's emphasis on the distinction between "variations" and "modifications." Variations were "deviations which, apparently, do not depend on the external environment, but occur as a result of the tendency to variation inherent to all living beings." Modifications resulted from the direct action of the environment.[66] In *The Flora of Eastern European Russia*, Korzhinskii attended closely to deviant forms and consistently attempted to discern if these were variations or modifications. He did so with the explicit intention of preparing a treatise on evolutionary theory.[67]

His budding ideas about evolution were particularly evident in his discussion of *Anemone patens*. (*Anemone* is a genus of the buttercup order.) This plant, he observed, grew abundantly in southern Siberia under the most varied conditions. As one traveled westward, its habitat became more limited and sharply defined. Korzhinskii concluded that southern Siberia was its center of distribution and observed that it was also the place in which *A. patens* exhibited the greatest variability.

> Numerous variations arise in Europe also, but there they do not find favorable conditions for their development and further consolidation: they either perish immediately in the struggle for existence or have only a local, limited distribution. But in southwest Siberia, where *A. patens* grows especially abundantly and is not subject, apparently, to such a terrible struggle for existence, newly arisen variations find entirely propitious circumstances for their further development. They take shape and spread out, begin to crowd one another and migrate, one after the other, in various directions.[68]

Korzhinskii observed that the oldest of these forms had consequently been pushed far to the west; it was rarely found in the original center of its distribution. In contrast, one young form was found primarily in this original, southeastern habitat, and there numerous recent variations were evident. These new forms were still volatile and unsettled. They were not yet independent races but, rather, "races *in statu nascendi*." Such observations, Korzhinskii added, "can yield very interesting theoretical conclusions concerning the very process of the origin of races."[69]

In this discussion the struggle for existence remained the key factor in geographical distribution, but it was also an obstacle to evolution. Where struggle was intense, new, unstable variations generally perished. Where conditions were more favorable, newly-arisen variations were able to consolidate themselves and multiply. The increasing population of these consolidated forms had a benign result with little evolutionary significance—migration.

Four years later, in his *Sketches of the Plant Life of Turkestan* (1896), Korzhinskii observed that a harsh struggle for existence also interfered with the progressive development of human beings. This book resulted from his efforts, on behalf of the Academy of Sciences and the Ministry of State Domains, to ascertain Turkestan's suitability for colonization. Korzhinskii reported that conditions there were much more suitable for humans than were those of central Russia. Favored by the hot climate and abundant moisture, the local population was able to work at a slow, gradual pace. They had no need of the *strada*, the period of frenzied labor during harvest time that was a permanent feature of Russian peasant life. "These propitious external conditions have created the easygoing, sociable character of the Sarts, this lively and intellectually developed people." Neither tyranny nor other perverse historical fates had blighted their character, and all aspects of their social relations were characterized by a gentle communality.[70]

In January 1899 Korzhinskii presented a short abstract of his theory of heterogenesis to the St. Petersburg Academy of Sciences. One year later the first part of a longer work, "Heterogenesis and Evolution: Toward a Theory of the Origin of Species," was published in Russian and German. Intended to establish the factual basis for his theory, it drew on his own observations and those of foreign authors to demonstrate the prevalence of discontinuous, heterogenetic variations in cultivated plants. The second section, which was to discuss the "enormous evolutionary role" of these variations, was not completed due to Korzhinskii's death at age thirty-nine.[71]

Darwin had been aware of sudden deviations from the parental type, Korzhinskii recalled, but had thought them of little significance. Darwinists had followed his lead and, absorbed in theoretical discussions, had added little to knowledge of these forms.[72] In 1864 A. Kolliker had argued that these variations generated new species by a process he termed "heterogenesis." Yet Kolliker's argument, poorly presented and weakly substantiated, had made little impact.[73]

Korzhinskii claimed that during his expeditions he had searched unsuccessfully for evidence of gradual transmutation and had finally concluded that Darwin's theory was incorrect. He saw many heterogenetic variations among plants, but it was impossible to examine the actual emergence of new species in nature. He had, therefore, turned his attention to domesticated plants. Here again he found Darwin mistaken. Not one horticulturalist "had ever developed new races by working with individual characteristics nor observed the 'accumulation' of them." Rather, all new forms emerged through "sudden deviations from pure species or hybrid forms."[74]

The "essence of heterogenesis" was that the birth of a single, distinctive individual gave rise to a "heterogenetic race." These deviant forms were not preadaptive. Like many nonheterogenetic variations, they originated, independent of external conditions, "in some internal processes, in some changes of the egg cell, the essence of which is currently beyond our understanding."[75] He speculated that this transpired in the pollen or seed bud sometime after fertilization.[76]

Korzhinskii offered a table comparing his theory with Darwin's (which he labeled "the theory of transmutation"). As it is so revealing of his case for heterogenesis, it is reproduced here in full:

According to the Theory of Transmutation	*According to the Theory of Heterogenesis*
1. All organisms have the capacity of variation in part as a consequence of internal causes, in part from the influence of conditions, use or disuse of organs, etc. This variability is constantly manifested in the form of small and unnoticeable individual distinctions.	Inherent to all organisms is variability, as an internal basis of their character, not dependent on external conditions. This variability, inhibited by heredity, usually remains hidden, but from time to time is manifested in sudden deviations.
2. As a consequence of the struggle for existence and selection those individual characteristics which are useful are consolidated and summarized, the useless ones disappear. In this manner all the characteristics and qualities of a species, being the result of continuous selection, must correspond to the external environment and serve the goals of the organism.	These sudden deviations can, under propitious conditions, form the beginning of stable races. Their characteristics, arising independent of external conditions, sometimes turn out to be useful for the organism, but can also not correspond to the external environment.
3. By force of continual selection and accumulation of characteristics all species continually change and gradually almost overflow into new ones, not departing during this from their normal physiological state.	All species, once having been formed, remain unchanging, but occasionally new forms are chipped off by heterogenesis. Such newly arisen forms, as a consequence of the disturbance of heredity, have a shaky constitution, which is reflected in lowered fertility and often in a general weakness of the organism. Turning into stable races, the new forms gradually restore their constitution.
4. This process can occur anywhere and under any conditions. The harsher the external conditions and more powerful the struggle for existence, the more energetic is the action of selection and the quicker is the origin of new forms.	It follows from the above that the origin of new forms can only occur under conditions favorable for the existence of the species; and the more favorable they are, that is, the weaker is the struggle for existence, the more energetic is their development. Under harsh external conditions new forms do not arise or, having arisen, very quickly perish.

According to the Theory of Transmutation	*According to the Theory of Heterogenesis*
5. In this manner, the struggle for existence and selection, which is generated by it, is the major factor of evolution.	The struggle for existence and selection is a factor limiting forms and suppressing further variations, but in no case facilitates the emergence of a new form. It is a principle inimical to evolution.
6. If there were no struggle for existence, there would be no selection and survival of the strongest, then there would be no development, no improvement whatsoever, since adapted forms would have no advantage over the others; as a consequence of interbreeding with these others the former would not accumulate useful characteristics.	If there were no struggle for existence newly arising forms would not perish. The world of organisms would grow into a mighty tree, all branches of which would flower; and the most distant species, which are now isolated, would be linked with all the rest by intermediate forms.
7. So-called progress in nature, or the improvement of organisms, is merely a more complex and complete adaptation to external conditions. This is attained by purely mechanical means of selection and the accumulation of characteristics useful in the given external conditions.	Adaptation, appearing as a consequence of the struggle for existence, is not at all a synonym for progress, since the higher, more complete forms are far from always more adapted to external conditions than are lower ones. It is impossible to explain the evolution of organisms by a purely mechanical path. In order to understand the origin of higher forms from lower ones it is necessary to allow for the existence in organisms of a special tendency to progress, either connected to or identical with a tendency to variation, and which leads organisms to improve, insofar as external conditions permit.[77]

A glance at these two columns reveals that Korzhinskii considered an important advantage of his theory to be its denial of a creative evolutionary role to the intertwined processes of the struggle for existence and natural selection. He contrasted Darwin's view that harsh struggle generated evolutionary progress (in the form of adaptation to external conditions) to his own belief that such struggle played only a negative role, hindering evolutionary progress (in the form of "higher, more complete" life forms) and pruning the rich tree of evolution that would flower fully in its absence.

Hunger, struggle, and death were not, therefore, necessary to evolutionary progress. If they were, one would expect new species to arise most often along the outermost boundaries of their geographical range, where the physical obstacles to their existence were most severe. But just the opposite was true. The variety of forms became progressively richer as one approached the center of a species' habitat.[78]

For Korzhinskii favorable conditions (that is, the absence of a severe struggle for existence) were necessary for the emergence of new species. This was the case because, first, such conditions provided the "surplus of life energy" necessary for variability. Second, the first generation of a heterogenetic form was physically weak, with an unstable hereditary constitution and a low rate of reproduction. It would vanish without a trace if subjected to a harsh struggle.[79]

> Therefore, the more favorable the external conditions, and, consequently, the weaker the struggle for existence, the greater the chances for the development of new races from a given species.... In the presence of harsh conditions and an intensive struggle for existence heterogenetic varieties are not formed, and if they do form they perish very quickly, leaving only the typical form.[80]

Heterogenesis also explained phenomena that perplexed Darwinists. The absence of transitional links between fossil forms attested to the saltatory nature of evolutionary change. The stability of many species under the most varied circumstances demonstrated that evolution resulted, not from the survival of favored individuals in the struggle for existence, but rather from the random appearance of heterogenetic forms.

Korzhinskii emphasized the distinction between adaptation and progress. Natural selection most assuredly played a role in adaptation; it acted on consolidated heterogenetic forms and on smaller variations, but was unable to produce new forms. So in an article of 1898 Korzhinskii attributed cleistogamy (the appearance of a flower that does not open and is self-pollinating) and all other adaptive phenomena to "variations and selection."[81] Yet these factors played no role in the origin of new forms or in evolutionary progress.

The superiority of the theory of heterogenesis was "especially striking" when "we transfer our conclusions to the world of human relations." Everybody knew that in human society "hunger and need do not lead to progress" and that "adaptation does not at all signify improvement." Individuals and entire peoples remained ignorant and backward if preoccupied with the need to acquire their daily bread. Many great intellects had been physically weak and sickly, and would surely have perished had their lives been too trying. Nor were great cultural figures, who constituted "the pride of humanity," always well adapted to their environment—sometimes they remained entirely unrecognized by their contemporaries.[82] Conversely, adaptation to one's environment often attested to a weakness, rather than strength, of character.

The message of Darwin's "cruel theory," that human progress required the annihilation of the weak by the strong, thus violated common knowledge. The theory of heterogenesis, on the other hand, taught that initially unstable and vulnerable mutant forms were the principal agents of a species' progress. Therefore, for the development of new forms and, consequently, for progress, it is necessary not only that the strong live, but that the weak also survive, strengthening their organism and preserving their characteristics, which are perhaps useless at present but have significance for the future. It teaches us, consequently, the broad tolerance of the well-known aphorism: live and let live.[83]

Acceptance of the theory of heterogenesis would not, unhappily, change the course of human events. The harsh struggle that characterized human life would continue, taking the form of "'the peaceful' competition of citizens of one state," war between nations, and the oppression of one people by another. "The strong triumph everywhere and the weak perish or drag out a miserable existence." Yet heterogenesis did counsel a morality superior to that of the Darwinists. In opposing the veneration of struggle, it taught the following:

> In the world of man, as in the entire organic world, progress does not consist in adaptation and victory in the struggle for existence, but is a result of an internal principle, the consequence of striving for the ideals of truth, kindness and beauty, which are profoundly rooted in the soul of man and which constitute, perhaps, but a partial manifestation of that tendency to progress which is inherent to life itself.[84]

Korzhinskii's emphasis on the evolutionary role of discontinuous mutations was praised by I. I. Mechnikov and others, but it struck Timiriazev as simply another attempt to refute Darwin's theory. In 1905 he commented sarcastically that, in return for a state subsidy, Korzhinskii had adapted to official ideology and had "quickly transformed himself into a militant anti-Darwinist."[85] Timiriazev's polemical point, of course, was that Korzhinskii's mutation theory was itself the product, not of any internal, independent, heterogenetic process, but rather of the most vulgar and unprincipled adaptation to the tsarist environment. The botanist A. Ia. Gordiagin rendered a sharply contrasting verdict; was he praising Korzhinskii himself as a heterogenetic form? "Regardess of where the current originated or where it was going," Gordiagin recalled, "Korzhinskii was constitutionally incapable of adapting his convictions to it."[86]

CHAPTER 5

Mechnikov, Darwinism, and the Phagocytic Theory

> Compelling facts from the real life of organic nature do not agree with the basic propositions of the selection theory.[1] (I. I. Mechnikov, 1876)
>
> The phagocytic doctrine developed entirely on the basis of Darwin's evolutionary theory.[2] (I. I. Mechnikov, 1895)

It may seem improbable to suggest that the career of a Nobel Prize-winning pathologist who spent the last twenty-eight years of his life in Paris at the Pasteur Institute could illustrate the characteristic Russian response to Malthus and Darwin. Yet the life of Il'ia Il'ich Mechnikov (1845–1916) contained many such improbabilities. For one thing, this winner of the Nobel Prize in Physiology or Medicine had neither physiological nor medical training. His early career was devoted almost entirely to invertebrate zoology and embryology; the years immediately prior to his formulation of the phagocytic theory found him preoccupied with two topics seemingly distant from pathology: digestion among metazoans and the beet weevil infestations in southern Russia. For another, this advocate of an all-encompassing scientific worldview advanced a phagocytic theory that was branded by many of the leading pathologists of his day as a teleological throwback to a prescientific past. Finally, in his early years, when Darwin's theory was all the rage among materialist youths of his ilk, Mechnikov was brashly critical of it; whereas by 1910, when that theory had been largely discredited in scientific circles, especially in his adopted home of France, he was effulgent in its praise.

How did a zoologist interested in agricultural pests and sharply critical of Darwin's theory come to enunciate a general theory of immunity, to defend it against the leading pathologists of his day, to win a Nobel Prize, and, from his place of honor at the Pasteur Institute, to solemnly declare that he owed it all to Darwin? The answer, I think, indeed lies in Mechnikov's typically Russian reaction to Darwin's Malthusian metaphor. His analysis of the struggle for existence was a guiding thread in his scientific life, linking his studies of zoology to his insight into pathology and explaining other paradoxes in the life of this prolific and creative figure.[3]

Scientist in Search of a Rational Worldview

Il'ia Mechnikov was the youngest of five children in the family of a small landowner in Kharkov province, and was one of three to attain some celebrity. One brother, Lev, became a well-known member of Russia's radical community, an adjutant to Garibaldi, and an accomplished geographer. Another, Ivan, became president of the Kiev House of Justice and an acquaintance of Lev Tolstoy, who described his final days in "The Death of Ivan Il'ich."[4]

In 1856 Il'ia entered a local lycée, where the intellectual winds of the 1860s were already sweeping through the student body. There he read Buckle's *History of Civilization in England* and eagerly embraced the argument that science was the chief source of social progress. He founded a science club, studied German in order to read Feuerbach, Vogt, Moleschott, and Büchner, and embraced atheism so emphatically that his schoolmates awarded him the nickname "*boga nyet*" ("there is no god").[5]

His circle at the lycée also read Herzen's journal *Kolokol* (*The Bell*) and other forbidden literature, but unlike many youth of his day, Il'ia dismissed political activity as a futile and superficial distraction. While studying in Europe in the 1860s, he met the famous radicals Alexander Herzen and Mikhail Bakunin through his brother Lev; although he enjoyed their company, his glimpses of political life only confirmed his early conviction that "science was immeasurably superior to politics."[6] Throughout his life he retained the faith in science's social mission, the materialist philosophy, and the disdain for political activity that he developed during his school years.[7]

These were also years of scientific achievement for the intense and precocious young man. He collaborated with the future zoologist V. V. Zalenskii (1847–1918) on a Russian translation of William Grove's *On the Correlation of Physical Forces*, studied Heinrich Bronn's *Classes and Orders of the Animal Kingdom*, and decided that the animal world could best be understood by examining its simplest forms. He was fifteen years old when his first article on infusoria was accepted by the prestigious *Bulletin of the Moscow Society of Naturalists*.[8]

On graduation Mechnikov traveled to Leipzig, Berlin, and Würzburg intent on enrolling in a German university. A series of mishaps and his own mercurial temper led him instead to return to Russia and enter Kharkov University, where he studied from 1862 until 1864. Under the direction of physiologist I. P. Shchelkov (1833–1909), he continued his work on lower organisms, studying a ciliated infusorium in order to determine whether its stalk had a muscular character. In an article published in *Müller's Arkhiv* in 1863, he argued that it did not. This conclusion was harshly criticized by the prominent German physiologist Wilhelm Kühne, and Mechnikov responded in kind.

By this time Russia's intelligentsia buzzed with rumors about the wunderkind of Kharkov University who had already mastered the microscope, published scholarly articles, and refuted the criticism of a senior western European scientist.[9] Mechnikov also wrote several characteristically combative book reviews during these years. In one he criticized a geology text written by a member of his own Kharkov University faculty. In another, written at age eighteen and submitted for publication in Dostoyevski's journal *Vremia* (*Time*), he identified serious weaknesses in a work that he had purchased in Leipzig—Charles Darwin's *Origin of Species*.[10]

A commitment to evolutionary theory underlay Mechnikov's choice of research topics. Organisms that "had found no place in definite animal or vegetable orders," he reasoned, might "serve as a bond between these orders and elucidate their genetic relationship."[11] So he examined singular freshwater forms that resembled Rotifera in some ways and worms of the nematode group in others. In 1864 he pursued similar studies on the island of Heligoland in the Baltic Sea. Traveling to Giessen to present this work to the Congress of Russian Naturalists and Physicians, Mechnikov met the zoologist Karl Leuckart. In 1865 the tsarist government awarded him a scholarship to study with Leuckart for two years, but this apprenticeship proved short-lived. Returning to Giessen from a vacation with his brother in Geneva, Mechnikov discovered that in his absence Leuckart had published, under his own name, an account of the life cycle of Nematoda that Mechnikov recognized as his own. Leuckart denied Mechnikov's charge of plagiarism, and there followed an angry public exchange.

During his brief stay in Giessen, Mechnikov read Fritz Müller's *Für Darwin*, which convinced him that "the key to animal evolution and genealogy should be sought in the most primitive stages, in those simple phases of development where no secondary element has yet been introduced from external conditions."[12] Committed to a research program in comparative embryology, he traveled to Naples, where he began an enduring collaboration with his friend and fellow countryman A. O. Kovalevskii.

Returning to Russia in 1867, Mechnikov quickly earned both a master's and a doctoral degree in zoology from St. Petersburg University. After one year as *dozent* there, he became professor of zoology and comparative anatomy at the recently founded Novorossisk University in Odessa, where he remained from 1870 through 1882. During these years his collaboration with Kovalevskii on comparative embryology and invertebrate zoology earned each a solid scientific reputation.[13]

Despite scientific and professional successes, Mechnikov was not a happy man in these years. His pessimism and dark mood permeated several essays of the 1870s in which he emphasized the fundamental "disharmonies" that afflicted human nature—that is, the contradictions within the human organism as well as those between it, nature, and society.[14] In "The Age of Marriage" (1871) he identified a basic disharmony between the age at which individuals first experience sexual urges and that at which they attain the social maturity necessary for married life. In "The Struggle for Existence in a Broad Sense" (1878), he observed that conflict did not favor "the higher representatives of mankind," but, rather, practical men unrestrained by moral considerations.

Mechnikov attempted suicide at least twice between 1873 and 1881. In 1873 his first wife, Beketov's niece, died after a long illness. Distraught, emotionally exhausted, and convinced that the gradual failure of his eyesight would deprive him of the pleasures of scientific work, he took an overdose of morphine. He vomited, however, which saved his life and left him in a state of torpor. His later account of this episode, related to his second wife and recounted in her adoring biography, captures something of his distinctive spirit:

He said to himself that only a grave illness could save him, either by ending in death or by awaking the vital instinct in him. In order to attain his object, he took a very hot bath and then exposed himself to cold. As he was coming back by the Rhone bridge, he suddenly saw a cloud of winged insects flying around the flame of a lantern. They were Phryganidae, but in the distance he took them for Ephemeridae, and the sight of them suggested the following reflection: "How can the theory of natural selection be applied to these insects? They do not feed and only live a few hours; they are therefore not subject to the struggle for existence, they do not have time to adapt themselves to surrounding conditions." His thoughts turned towards Science; he was saved; the link with life was re-established.[15]

For a year or two afterwards, Mechnikov pondered the sudden death of the Ephemeridae (mayfly), hoping thereby to understand the dynamics of the struggle for existence, but he was ultimately stymied by the difficulty of studying physiological processes in an insect with such a short life span.[16]

His research interests shifted sharply in the late 1870s, when he decided that the "exclusively morphological-embryological path" could shed but little light on evolution. He resolved instead to investigate the phylogenetic development of a physiological system, and chose the digestive apparatus in Metazoa.[17]

Mechnikov's attention was directed to an immediately practical task as a result of his frequent visits to his father-in-law's estate in southern Russia. This agricultural region was suffering from pest infestations, and Mechnikov, having previously studied various parasites, developed a form of biological warfare against the beet weevil. His first work on the subject was published by the local entomological commission in 1878, and he continued this research through the mid-1880s.

With the assassination of Tsar Alexander II in 1881 and the subsequent wave of political repression, Mechnikov found Novorossisk University increasingly uncongenial. The tightening of state control over the university, factional disputes among the faculty, and constant student demonstrations disrupted his teaching and research. He had resolved to leave the university for a position as entomologist for the Poltava *zemstvo* (local government) when his father-in-law's death left him financially independent. Resigning his faculty position in May 1882, he departed for Messina to continue his research on digestion in lower organisms.[18]

The following year, "the great event of my scientific life took place"—the formulation of the phagocytic theory. Drawing on his studies of embryology, parasitology, and digestion, Mechnikov concluded that inflammation reflected a struggle between a parasite and the body's white cells. The same cells responsible for intracellular digestion in lower organisms, he reasoned, constituted the defensive forces of the host in higher organisms.

Mechnikov's remaining years were devoted to the defense and extension of the phagocytic theory. He responded to critics in *Lectures on the Comparative Pathology of Inflammation* (1892). Beginning with several papers on cholera in 1893–1897, he applied the phagocytic theory to a range of diseases, including plague, tuberculosis, syphilis, and typhoid. He updated and extended his general argument in *Immunity in Infective Diseases* (1901) and extended it still further in *The Nature of Man* (1902) and *The Prolongation of Life* (1907), in which works he offered an analysis of aging and an all-embracing philosophy of life.

Most of this work was conducted in France, his adopted country. The final decision to emigrate was precipitated by local reaction to the Odessa Bacteriological

Station, which Mechnikov, N. F. Gamaleia, and Ia. Iu. Bardakh had founded in 1886. Here were trained such future leaders of Russian bacteriology and epidemiology as P. N. Diatropov, L. A. Tarasevich, and D. K. Zabolotnyi. Local physicians, journalists, and politicians denounced the station as a threat to the community, however, and it was temporarily closed in 1887 for its reputed role in an epizootic of anthrax.[19] Convinced that the obstacles to scientific work in Russia were insurmountable, Mechnikov scouted for an institutional home in western Europe. Robert Koch greeted him coolly, but Louis Pasteur thought the phagocytic theory might provide a much-needed explanation of immunity and offered Mechnikov a laboratory in his new institute.[20] Mechnikov moved to Paris in 1888 and remained at the Pasteur Institute for the remainder of his life. He was elected to the French Academy of Sciences in 1904, and in 1908 shared the Nobel Prize in Physiology or Medicine with Paul Ehrlich, a leading humoralist opponent of the phagocytic theory. He left Paris four times to visit Russia: in 1894 to report on measures against cholera, in 1897 to attend the Twelfth International Congress of Physicians in Moscow, in 1909 to visit former colleagues and meet with Lev Tolstoy, and in 1911 to study tuberculosis among the Kalmuks of the Astrakhan steppe.

The phagocytic theory proved to be a vehicle for Mechnikov's transformation from pessimist to optimist. Pessimism, he concluded, originated in the recognition of a profound disharmony in human nature: the inevitability of physical and mental decay in old age and the impossibility of attaining a "natural death." That is, unlike such creatures as the Ephemeridae, humans died while the life instinct was still strong. Some sought to assuage their fears with "childish" religious conceptions and various idealist philosophical systems, but only science could explain and alleviate the offending disharmony itself. Premature deterioration and death, in Mechnikov's view, resulted from the conquest of healthy organic elements by the harmful microbes that flourished in the large intestine.[21] By intervening in this struggle (Mechnikov recommended a daily dose of lactic acid), science would eventually extend the human lifetime, bringing it into harmony with the life instinct and allowing humans a natural, "orthobiotic" death.

"Science is my politics," he often declared, and he wrote enthusiastically of the social consequences of scientific progress in general and the prolongation of life in particular. Mechnikov found unacceptable both existing social inequalities and the restrictions on individual liberty that he feared would result from socialism. The accumulation of positive knowledge, he argued, would itself result in "a great levelling of human fortunes" and so provide the basis for the solution of all social problems.[22] Furthermore, science's prolongation of life would allow vigorous and experienced old men to run human affairs, thus avoiding the problems caused by young and impetuous leaders.[23] In 1907, stung by the accusation that his scientific utopia was too individualistic, he went further and projected a ruling technocratic elite:

> It is easily intelligible that in the new conditions such modern idols as universal suffrage, public opinion and the referendum, in which the ignorant masses are called on to decide questions which demand varied and profound knowledge, will last no longer than the old idols. The progress of human knowledge will bring about

the replacement of such institutions by others, in which applied morality will be controlled by the really competent persons.[24]

Mechnikov was, then, understandably disturbed by his perception at the turn of the century that science's prestige was suffering amid the vogue of philosophical idealism. Speaking to the Twelfth International Congress of Physicians in 1897, he took issue with those who proclaimed the "bankruptcy of science."[25] In *The Nature of Man* he wrote of the "open war" between science and the superstitions offered by religious dogma and philosophical idealism.[26] For Mechnikov, Bergson and other philosophers threatened to replace science with rationalized religious dogma while agnostics such as Du Bois Reymond abandoned the field, tacitly encouraging the false belief that science "has robbed mankind of the consolations of religion without being able to replace them with anything more exact or enduring."[27] In truth, if humans were ever to solve their problems, they "must be persuaded that science is all-powerful and that the deeply rooted existing superstitions are pernicious."[28]

World War I provided a barbaric riposte to Mechnikov's faith in the civilizing influence of positive knowledge. He professed great disappointment in the behavior of scientifically advanced Germany and watched sadly as young men left his laboratory for the front, where several perished. He died in 1916, convinced that his healthy cells had lost a struggle for existence with harmful intestinal flora and concerned that his death might discredit his ideas on the prolongation of life.

Essays on the Selection Theory, 1863–1878

Mechnikov was eighteen years old in 1863 when he first wrote on Darwin's theory. Although unpublished during his lifetime—the state suppressed the journal for which it was intended—his 1863 review reveals his youthful enthusiasm for evolutionism and his reflexive antipathy for Malthus.

The major weakness in the *Origin*, he contended, was the author's "generalization of the Malthusian law and his attribution of special significance to the principles of natural selection and extinction."[29] This Mechnikov attributed to Darwin's "poverty of facts" and his "too superficial view of the influence of external conditions on the organism, which, of course, is the main fact of the organization of life."[30]

This superficiality was manifest in Darwin's view that overpopulation was the most important source of the struggle for existence, whereas high fertility was actually more often the consequence than the cause of struggle. Furthermore, an organism's rate of reproduction was not fixed, but varied greatly under the influence of volatile environmental conditions. Referring to Beketov's law of fertility, Mechnikov commented that "it is well known that the greater the dangers to which a given organism is subject, the greater its means for reproduction."[31]

Darwin also violated common sense by his assertion that the struggle for existence was most intense among closely related organisms:

> This opinion is completely untrue: first, because food (organisms) also . . . increases at the same rate as the organisms making use of it; they, in turn, serve as nutriment

for other beings. Second, ... as everyone knows shared dangers and obstacles do not stimulate struggle between the individuals subject to them, but, on the contrary, impel them to unite together in one society and to resist these obstacles with joint, more reliable forces. It even seems to me that a similarity in organization facilitates the absence of struggle between beings.[32]

For Mechnikov, Darwin's relationship to Malthus was contradictory. On the one hand, Darwin relied on and invoked Malthus, which distorted his view of nature and weakened his selection theory. On the other, Darwin actually refuted Malthus's central proposition, since the naturalist contended that *all* organisms, and consequently the food supply for humans, tended to multiply at a geometrical rate.[33] Despite the Malthusian weaknesses in Darwin's argument, Mechnikov predicted a "great future" for evolutionary theory itself.[34]

His position changed somewhat in the 1870s. He remained adamantly opposed to Darwin's Malthusianism but made no mention of mutual aid; nor did he simply reject the selection theory as a fallacious "generalization of the Malthusian law." Rather, he accepted the struggle for existence as a useful concept and attempted to analyze its dynamics, purge it of Malthusian distortions, and define its limits as a factor in evolution. His harshest criticism was directed, not at Darwin, but at "Darwinists." In their insistence that overpopulation, intraspecific competition, and natural selection were "the only basis of all phenomena of the organic world," these dogmatists ignored the many phenomena that "cannot be reconciled with the now dominant doctrine of the origin of species."[35]

The difference between the founders of the selection theory and their successors emerged clearly in their approach to the struggle for existence. Darwin and Wallace had simply appropriated this concept from Malthus; in their works the struggle for existence lacked precision and explanatory power, remaining simply a metaphorical expression for "the sum of phenomena resulting in either survival or death."[36] Their successors were more precise but less accurate. They tended to interpret the struggle for existence more narrowly, separating "competition between the most closely related organisms from the entire sum of phenomena of the struggle for existence."[37]

G. Seidlitz, the Darwinist zoologist at Dorpat University, had even suggested that the term "struggle for existence" be replaced by "competition of like with like" to express more precisely the emphasis of the selection theory. Seidlitz's formulation epitomized the "untenable extremes" implicit in Darwinists' preoccupation with intraspecific competition:

> From this point of view one must say that, for example, when the black rat is supplanted by another species the actual struggle for existence occurs not between the two species but among the individuals of each species, that brown rats compete with one another over who can best oust the black rats and take their place. Similarly, in a war between two peoples, the struggle for existence will not consist in the war itself, but rather in the rivalry between individuals of each people for services and distinction.[38]

Such dogmatism was useless to the naturalist. The rich complexity of natural relations was more accurately reflected in Darwin's original formulation, which required clarification rather than restriction. Mechnikov suggested the following classification of the struggle for existence: (1) competition (*konkurentsia*) between members of the same species; (2) competition between members of different species; (3) struggle (*bor'ba*) between members of different species; and (4) struggle between organisms and environmental conditions.[39]

The interaction of these different aspects determined the results of conflict in the natural world. Illustrating his point with examples from the life of ants, bees, and other organisms, Mechnikov argued that the character of intraspecific relations—the degree to which they were dominated by different types of struggle or by cooperation—was quite fluid and dependent on environmental circumstances. The evolutionary consequences of different aspects of the struggle for existence were similarly varied and could be determined only by examination of specific cases.[40] Mechnikov acknowledged that "from the Darwinist perspective," intraspecific struggle and competition were the only aspects of the struggle for existence that could actually produce new species. He insisted, however, that these did not generate new forms unless stimulated by interspecific conflict.

This contention was related to his most uncompromising criticism of Darwin's theory—his rejection of the Malthusian assertion that overpopulation stimulated the struggle for existence and evolution. As was apparent from a careful reading of the *Origin*, this Malthusian contention had trapped Darwin within a multitude of paradoxes and self-contradictions. For example, freshwater plants were known both to reproduce rapidly and to exhibit a relative paucity of new species. This would not be the case if overpopulation generated intraspecific competition and if this competition, in turn, resulted in the evolution of new forms. Mechnikov cited Darwin's explanation for the relative stability of freshwater plants: "All freshwater basins, taken together, make a small area compared with that of the sea or of the land; and consequently the competition between freshwater productions will have been less severe than elsewhere; new forms will have been more slowly exterminated."[41] This passage appeared in a section of the *Origin* in which Darwin discussed the advantages of a large area for the generation of new forms. Species arising in large areas, Darwin explained, benefitted from a sizable pool of potentially favorable variations and from their adaptation to complex conditions of life. As a result they were able to endure for long periods of time and in varying environments. Mechnikov commented:

> Here Malthus's law, which constitutes such an important foundation of Darwinism, is forgotten: limited space should most of all strongly facilitate overpopulation (as we do in fact see in reality) and overpopulation should stimulate the formation of new characteristics and the decline of old species in favor of new ones. If this is not so, if competition between different organisms is necessary for the formation of new species, then clearly the factor of overpopulation falls to the background in transformism.[42]

Similarly, Darwin explained the relative stability of such lower forms as amphioxus as a consequence of their life in isolation from other organisms. Here again, Darwin contradicted himself and "forgot" Malthus. "From a truly Darwinist point of view," Mechnikov insisted, the most important competition ensued among individuals of the same species, amphioxus itself. "If it lives in isolation it should multiply without hindrance in a geometrical progression, and this circumstance should in itself lead to variations."[43] Even when an isolated species did multiply mightily and densely populate an area, the results rarely justified the Darwinists' expectations.

> Salt-water pools present a very sharp instance of a closed space with a specific population. Many are populated almost exclusively by one species of animals, specifically the cancroid *Artemia salina*, which is found in enormous number. With its rapid and constant multiplication this animal often attains the limit of population density and so should be subject to a strong struggle for existence. This information would cause one to suppose the heightened activity of natural selection in general and the law of the divergence of characteristics in particular. But it turns out that *A. salina* is among the most constant of species.[44]

Embryological evidence also refuted Darwin's contention that overpopulation and intraspecific competition produced a divergence of characters and the evolution of new species. If they did, and if, as Darwin contended, natural selection operated throughout the life of an organism, then one would expect more variations among the young than among adults, since the population was larger in the earlier stages of life. This, however, was not the case. Larvae did not exhibit more sharply defined species characteristics than did adult forms, not even among Ephemeridae, whose larvae lived for an entire year, compared with the adult form's life of only an hour or so.[45]

Mechnikov also elaborated his earlier argument that Darwin, like Malthus, dealt too abstractly with the dynamics of fertility: "Darwin isolates the phenomenon of high reproduction too sharply and places it at the basis of struggle and selection, when it is itself subject to change and to the influence of natural selection."[46] Rapid reproduction did sometimes stimulate struggle; much more frequently, however, it was itself an organism's response to an intense struggle for existence, and was often the determining factor in that struggle's outcome.[47] High fertility simply "does not have the great significance as a stimulant of struggle that Darwin ascribed to it."[48]

What, then, stimulated the struggle for existence? For Mechnikov the answer was simple: the efforts of organisms to satisfy their needs in the most direct and efficient manner possible.

> Attempting to acquire the greatest amount of nutriment in the most direct manner, bees conduct raids upon others and struggle among themselves. Ants commonly engage in bloody battles, in the course of which they capture slaves, etc. It is understandable, then, that one can explain the universality of the struggle for existence without accepting the view that the earth is overpopulated and that struggle is always generated by the extreme, urgent need for a piece of bread. One can affirm confidently that in many places on earth (for example, on oceanic islands) the animal population has not even approached its limit, which does not prevent the most numerous and varied manifestations of struggle.[49]

The struggle for existence was most severe, and most often led to the evolution of new forms, when the needs of different organisms brought them into conflict with one another. This "competition and struggle of different forms" played an "incomparably greater role" than overpopulation in the origin of new species.[50]

Human conflict had the same non-Malthusian origin. In his article "The Struggle for Existence in a General Sense" (1878) and in his unpublished notebooks on *The*

Descent of Man, Mechnikov attributed struggle among humans to the variety and infiniteness of human desires:

> Two tribes conflict, not because the means for survival are inadequate for their large numbers, but as a result of mutual hatred and the desire to enslave others; generally, from laziness and the desire to exploit another. Consider an example. On an island resides a certain number of people with plentiful resources for their existence. Still, there is a lazy person among them who will exploit others based on his simple desire to do nothing, in other words, to satisfy his desire for idleness. Being shrewder, he becomes richer and stronger, without working, than his industrious, but less shrewd neighbors. Here the source of inequality, and consequently of struggle, lies not in heightened reproduction but in human inclinations. Therefore, we see that even in a country where there are many sources of nourishment (even Russia) there appears a struggle conditioned by inequality.[51]

Conflict was indeed inherent to nature, including human society. It did not, however, arise from a mythical disproportion between population and food supply—who would claim that such a disproportion existed in Russia, where inequality and conflict nevertheless abounded?—but simply from contradictory aspirations and requirements within a given population. This was true among bees and ants, with their relatively simple drives, and among humans, who frequently desired wealth, power, fame, or idleness.[52]

Parasite as Paradigm

Mechnikov's analysis of the struggle for existence owed much to his years of research on parasites.[53] These he summarized in his "General Essay on the Life of Parasites" (1874), which explored the struggle for existence between parasite and host as an example of a confrontation between simpler and more complex species.

His essay posed the following central question: why do parasites attach themselves to the bodies of other organisms? What, in other words, stimulates the struggle for existence? Even a glance at the variety of circumstances under which parasitism occurs revealed that it was not "dependent upon the earth being too densely populated."[54] At any rate, "the view that every corner of the earth is teeming with life is positively untrue."[55] Parasites lived as they did because, quite simply, "their way of life has turned out to be very advantageous for the existence of individuals and for the leaving of progeny."[56]

Why, he then asked, are parasites so successful in the struggle for existence? Two factors were most important: their exceptionally high rate of reproduction and their physiological "hardiness."[57] Thus, when higher organisms "enter a struggle with the small parasites their antagonist is much more simply organized, but is by no means always weaker."[58] Mechnikov observed that parasites were frequently successful in their struggle with human organisms but conceded that the dynamics of this struggle involved medical questions beyond his expertise.[59]

Parasites' frequent victories against higher organisms demonstrated that natural selection did not always favor complexity of organization. Indeed, as Mechnikov pointed out in his "Essay on the Origin of Species" (1876), Darwin's examples of the struggle for existence rarely ended with the triumph of the more complex form.[60] Natural selection, therefore, could not explain "the progressive development of organisms." This basic tendency in nature, it seemed, "depends upon a special tendency to perfection [*stremlenie k sovershenstvovaniiu*]."[61]

Furthermore, the factors most important to the parasites' success in the struggle for existence—high fertility and hardiness of constitution—were "internal physiological characteristics" with little relationship to the systematic morphological features that defined species. Natural selection, then, could hardly explain speciation in such cases. The reverse was also true: "Species characteristics can be formed in cases where the action of natural selection is greatly limited." Darwin dismissed such cases as exceptional, "but actually they are widespread and powerful phenomena."[62]

So, while Darwin's establishment of a causal connection between the struggle for existence, natural selection, and speciation "must be recognized as a firm contribution to science," this connection was only an occasional one. A comprehensive theory of evolution would necessarily include other mechanisms, and these might well prove more important than natural selection.[63]

These insights into parasitism and the struggle for existence figured prominently in Mechnikov's efforts to combat the beet weevils (*Anisoplia austriaca*) of southern Russia. This work began in 1878 and continued through the early 1880s—that is, immediately before, during, and after his formulation of the phagocytic theory.[64]

In "Illnesses of the Larvae of the Beet Weevil" (1879), he proposed that directed interspecific struggle, rather than the prevalent mechanical means, was the best means of controlling the beet weevil. "The entire sum of scientifically acquired facts," he explained, "leads to the conclusion that lower parasitic organisms are actually the most powerful enemies of animals and therefore can be used to relieve man of harmful insects."[65]

Mechnikov's conviction that interspecific conflict, rather than overpopulation, underlay the struggle for existence played an important role in his reasoning about the beet weevil problem. He was struck by the intermittency of complaints about the beet weevil—these were registered in waves in 1857, 1862, 1875, and 1878. Why did the pest prosper in these years and not in others? The answer could not lie in the periodic exhaustion of its food supply—local landowners had brought in a harvest even in the worst years of infestation. He concluded, rather, that the beet weevil was locked in a seesaw battle with some unidentified foe.[66]

Mechnikov subsequently discovered that three different parasites afflicted the larvae of the beet weevil. In 1879-1880, encouraged by a local landowner (who also convinced him of the inefficacy of communal ownership of the land), he experimented with one of them, muscardine, and concluded that this was the most potent and practical weapon against the beet weevil.

Over the next few years Mechnikov continued these experiments and polemicized against opponents of his project.[67] He developed large quantities of muscardine in 1883-1884, but his proposal to use it on local farms was never implemented, perhaps because of an economic crisis in the sugar industry.[68] His preoccupation with parasitism and interspecific struggle, however, had already contributed to a theoretical insight of great importance for his career and for medical science.

Intracellular Digestion and the Phagocytic Theory

While working on the digestive physiology of metazoans in Messina in 1883, Mechnikov formulated the phagocytic theory of inflammation for which he would become internationally renowned. "A zoologist until then," he later recalled, "I suddenly became a pathologist."[69] Certainly this theory represented a sudden leap across disciplinary lines into unfamiliar territory. Yet this leap was prepared and negotiated by his confidence that zoology and pathology were linked by three subjects that he knew well: parasitism, evolutionism, and the dynamics of the struggle for existence.

When Mechnikov began his studies of digestion, it was well known that protozoans digested nutriments within their individual cells.[70] Interested in the evolution of digestive systems, he traced this phenomenon through more complex forms and quickly determined that intracellular digestion also characterized such lower metazoans as sponges and coelenterates. Even among higher metazoans with a digestive cavity, the actual digestive process occurred within the mesodermal cells that lined the cavity's walls.

In 1879, while working in Naples and Messina, Mechnikov established that the amoeboid mesodermal cells of the larvae of echinoderms and coelenterates retained their digestive capacity. This was interesting because such a capacity seemed redundant given the presence in these forms of a digestive tube (which developed from the endodermal germ layer). He also observed that during metamorphosis these amoeboid cells destroyed superfluous parts of the larval form and that they gathered within them any carmine bodies suspended in the surrounding solution.

He soon concluded that these amoeboid mesodermal cells, which were responsible in the simplest organisms for digestion, retained their digestive capacity even in higher organisms, where digestion was largely the function of the endoderm. Furthermore—and it was this perception that transformed him suddenly into a pathologist—he concluded that among higher forms this digestive capacity was turned to the defense of the organism against the invasion of foreign bodies.

Mechnikov twice described the circumstances that prompted this conclusion. Both descriptions were written years after the fact. One, intended for a scientific community skeptical of his conclusions, appeared in 1901; the other, for a popular audience, in 1908. The oft-quoted popular account was in the best tradition of "eureka" narratives.

> One day when the whole family had gone to a circus to see some extraordinary performing apes, I remained alone with my microscope, observing the life in the mobile cells of a transparent star-fish larva, when a new thought suddenly flashed across my brain. It struck me that similar cells might serve in the defence of the organisms against intruders. Feeling that there was in this something of surpassing

interest, I felt so excited that I began striding up and down the room and even went to the seashore in order to collect my thoughts. I said to myself that, if my supposition was true, a splinter introduced into the body of a star-fish larva, devoid of blood vessels or of a nervous system, should soon be surrounded by mobile cells as is to be observed in a man who runs a splinter into his finger. This was no sooner said than done. There was a small garden to our dwelling; ... I fetched from it a few rose thorns and introduced them at once under the skin of some beautiful starfish larvae as transparent as water. I was too excited to sleep that night in the expectation of the result of my experiment, and very early the next morning I ascertained that it had fully succeeded. That experiment formed the basis of the phagocyte theory, to the development of which I devoted the next twenty-five years of my life.[71]

As we shall see shortly, this account of the origins of the phagocytic theory would hardly have impressed a medical audience familiar with the literature on inflammation.

Mechnikov's other version, which he included in his *Immunity in Infective Diseases* (1901), stressed the logical continuities in his thought process. Here he revealed that before his departure for Messina he had read Julius Cohnheim's treatise on general pathology. Cohnheim's description of the movement of white blood cells had provoked Mechnikov to consider an analogy between the function of these white cells and the digestive function of the amoeboid mesodermal cells. "This reflection," he recalled in 1901, precipitated the critical experiment with the starfish larva.[72]

Interpreting his observations boldly, and doing so before witnessing even a single instance of spontaneous phagocytosis in higher organisms, Mechnikov concluded that the mobile white cells were engaged in interspecific conflict with the foreign bodies, and that this was the essence of inflammation throughout the organic world. In his first communication on the subject, "Investigations of Intracellular Digestion in Invertebrates" (1883), he described the struggle between the white cells and bacteria: "All these bacteria are hotly pursued by the mobile cells inside which one can observe bacteria in all stages of development and digestion. But the struggle is waged by both sides and dead mobile cells are, apparently, among the fallen; rays of escaping bacteria stream from them."[73]

In 1883 Mechnikov presented his theory to the Seventh Congress of Russian Naturalists and Physicians in a speech entitled "On the Curative Forces of the Organism." Although he knew little about medical science, and less about medical history, he confidently portrayed the phagocytic theory as the logical culmination of medical ideas from Hippocrates to Virchow.[74]

By this time he had developed the basic components of his theory: (1) inflammation is a therapeutic, rather than a pathological, phenomenon; (2) it reflects the active struggle of the white blood cells against foreign bodies; and (3) it is but one stage in an evolutionary continuum that begins with the struggle of unicellular organisms against their foes and includes the entire gamut of host-parasite relationships.

Mechnikov often expressed surprise and disappointment at the negative reaction to his theory by physicians and medical scientists. His wife later recalled that he was bewildered by "this obstinate opposition to a doctrine based on well-established facts, easily tested and observed throughout the whole animal kingdom."[75] Being but slightly acquainted with the existing literature about inflammation, Mechnikov was, of course, ill equipped to predict the medical community's response.[76] According to his wife, it was only after initial critical reactions that, much "grieved and pained," he resolved "to study the medical side of the question in order to prove on that ground that his theory was well-founded."[77] Mechnikov himself recalled in 1901 that his admittedly cursory review of pathologists' previous work had convinced him they would receive his theory warmly:

> I thought ... that the observations on absorption and leucocytes, which had been accumulating for years in pathological histology had sufficiently paved the way for a favourable reception to the idea that the amoeboid cells are defensive elements of the body capable of guaranteeing to it immunity and cure. In this I was mistaken. It was precisely the specialists in this branch of science who from the first manifested the most lively opposition to this theory.[78]

He soon discovered that this "lively opposition" was rooted less in factual discrepancies than in a deep-seated resistance to the kind of explanation that he offered. To appreciate this resistance, and to understand its implications for Mechnikov's attitude toward the selection theory, we must digress briefly.

Pathologists and Inflammation

Pathologists had for some time grappled with many of the "well-established facts" that, for Mechnikov, provided decisive evidence for the phagocytic theory. Mechnikov's contribution resided primarily in his interpretation of these facts, and in his insistence that infection was the struggle between two organisms and that inflammation was a defensive response to a parasite's attack upon the body. For Mechnikov this interpretation flowed logically from an understanding of parasitism, evolutionary theory, and the struggle for existence.[79] For well-versed pathologists, however, it smacked of teleological theories that had been long ago discredited.

L. J. Rather and other medical historians have documented the long and checkered history of the notion that inflammation was a therapeutic process reflecting the body's struggle against foreign invaders.[80] Many eighteenth-century thinkers, including Thomas Sydenham, George Stahl, and John Hunter, had propounded the notion of "laudable pus." William Addison spoke for at least a significant minority of English physicians when he wrote in 1868: "Nature has the action of inflammation in reserve, not for the purpose of vexing mankind and shortening life, but for the purpose of repair and healing—the cure of wounds and fractures, the discharge of dead parts and foreign bodies, and for elimination of unwholesome poisonous matters from the blood."[81] Nor was Mechnikov's blinding insight at Messina, his analogy between inflammation and an organism's reaction to its wounding by a thorn, a novel one within medical circles. Galen had appreciated this analogy some seventeen hundred years earlier, and van Helmont had speculated in the seventeenth century about a "metaphorical thorn" responsible for pleurisy.[82] The military metaphor was equally familiar to readers of the medical literature. During the first half of the nineteenth century, Carl Heinrich Schultz and many others had referred to the "defensive processes" of the body in "the battle against disease."[83]

In the decades prior to Mechnikov's work on inflammation, however, the notion of "laudable pus" and bodily defenses had been largely discarded by Continental pathologists. According to Rather, this age-old idea fell casualty to the reaction against romanticism, which was expressed in "a strong anti-teleological sentiment among scientifically oriented physicians."[84] Rudolf Virchow, who had himself previously employed the image of bodily "gendarmes," wrote in 1847 that scientific pathology had finally purged itself of such teleological conceptions:

> Suppuration is no longer a therapeutic effort of the organism to fill this or that breach, the pus corpuscles are no longer the gendarmes ordered by the police state to escort some stranger or other without a passport over the border; scar-tissue no longer constitutes the prison walls within which this stranger is enclosed when the prison-organism so pleases.[85]

It was, then, within a resolutely antiteleological framework that leading pathologists interpreted the "well-established facts" that Mechnikov, in his innocence, had expected to win acceptance for the phagocytic theory.

For example, Mechnikov thought the presence of bacteria within white blood cells was persuasive evidence for his theory, but this well-known fact lent itself to a variety of interpretations. Robert Koch, for instance, had noted the presence of the anthrax bacillus within the white cells of diseased animals (and Mechnikov had cited this as one "well-established fact" in favor of the phagocytic theory). For Koch, however, the white cell was a passive medium; bacteria penetrated it, multiplied within it, and used it as a means of transport throughout the body.[86] Rudolf Virchow had earlier warned Mechnikov that Koch was not alone, that most pathologists "believed that the leucocytes, far from destroying microbes, spread them by carrying them and by forming a medium favorable to their growth."[87]

The efforts of the renowned pathologist Julius Cohnheim to explain the migration of white cells toward a wound manifested this same reluctance to see them as active entities. In 1867, in the same work that Mechnikov consulted before his departure for Messina, Cohnheim described the passage of white blood corpuscles through the vascular wall to the site of irritation. He thought that this demonstrated the cells' active character but, wary of teleological reasoning, could not explain their migration to the precise location where they were needed. Simon Samuel provided an answer in 1870: infection increased the permeability of the vascular wall by causing chemical changes in tissue fluids. White cells, then, did not move actively toward a wound but simply seeped through an altered vascular wall. Cohnheim adopted this perspective in 1873. Fortified with this acceptable physicochemical explanation of the cells' motion, he explicitly rejected as teleological the idea that inflammation was an organism's reaction to pathogens. "Today," he commented, such views "enjoy little esteem."[88]

Mechnikov's initial communications about the phagocytic theory, then, offered pathologists very little new information. The migration of white cells toward an infection or wound, the presence of bacteria within these white cells, the similarities between white cells and freestanding amoebae, with their common capacity of intracellular digestion—all these phenomena had been well-known and widely scrutinized. His "conclusive experiment" with thorn and starfish must have seemed amateurish and naive to anybody familiar with the issues involved. Furthermore, in his initial communications Mechnikov relied on observations of lower organisms, making him an easy target for physicians and pathologists with broader experience.[89]

It is not surprising, then, that the phagocytic theory was roundly criticized. Criticism escalated after Mechnikov reached a broader audience with his article "Sur la lutte des cellules de l'organisme contre l'invasion des microbes" (1887), which appeared in the first volume of the prestigious *Annales de l'Institut Pasteur* in 1887.[90] No doubt the skepticism of German pathologists was further aroused by their French rival Pasteur's glowing reference to Mechnikov's "very original" and "fertile" theory.[91]

Many pathologists insisted that inflammation was not the benign, defensive process that Mechnikov claimed it to be, that the white cells were often the passive accomplices of infection and, most importantly for our purposes, that the phagocytic theory was teleological and vitalist. For example, in a speech to the 1891 London Conference on Hygiene and Demography, Emil Behring charged that Mechnikov's hypothesis relied on "secret forces of the live cell." In a letter to Mechnikov in 1892, he characterized the "struggle theory," with its reliance on "some vital activity," as "an excursion into the realm of metaphysical speculations."[92] William Henry Welch, professor of pathology at The Johns Hopkins University Medical School, also warned against the reintroduction of teleological reasoning under cover of the phagocytic theory. He cautioned against attributing "something in the nature of an intelligent foresight on the part of the participating cells" and suggested that the best means of avoiding this was the search for a "mechanical" explanation based on a knowledge of cellular properties.[93]

Opposition to the phagocytic theory crystallized around a "humoralist" alternative after George Nuttall's 1888 report on animal serums that were toxic for certain microorganisms. The debate between "humoralists" and "cellularists" continued until World War I.[94]

It is important to bear in mind that Mechnikov's critics did not deny that white cells sometimes played a role in disposing of pathogens. This was not, as we have seen, a novel idea. What they did deny, and what Mechnikov insisted upon as the core of his theory, was that "the essential and primary element in typical inflammation consists in a reaction of the phagocytes against a harmful agent"[95]—that is, that inflammation was the product of interspecific conflict. Throughout the debate Mechnikov attempted to incorporate humoralist findings within this framework. If he were claiming merely that phagocytes often played an important defensive role, he could have simply accepted the observation that white cells sometimes conducted the pathogen throughout the body. But he was claiming much more, so any view that relegated phagocytes to a passive rather than an active role and made them ancillary, rather than central, to the inflammatory reaction struck at the heart of his conception. Similarly, as his opponents accumulated evidence of the importance of humoral antibodies, Mechnikov attempted to incorporate it into the phagocytic theory by arguing that these antibodies were produced by phagocytes, were of relatively late evolutionary origin, and, in any case, served mainly to prepare the ground for the action of the white cells.[96] Again, this argument defended his "central dogma" that inflammation was essentially a conflict between two species.

This general point is evident from Rather's examination of one pathologist's reaction to the phagocytic theory. Ernst Ziegler's *Lehrbuch der Allgemeinen und Speciellen Pathologischen Anatomie und Pathogenese* was one of the medical works that had encouraged Mechnikov to believe that pathologists had accumulated a mass of evidence "fitted to facilitate the acceptation of the new hypothesis on inflammation and healing."[97] Here Ziegler discussed the movement of white cells and the damage caused by bacteria to vascular cells. He noted that "inflammation, aroused by bacteria, in the course of which a great number of living cells accumulate in the tissues, is very frequently able to suppress the invasion of bacteria."[98] The means by which inflammation suppressed bacteria was unclear, but when the "bacteria are killed" the infection healed.[99]

One can easily understand why Mechnikov found these passages encouraging, yet subsequent editions of Ziegler's *Lehrbuch* proved his optimism illusory. In the fourth edition (1885) Ziegler noted that Mechnikov had christened the well-known amoeboid cells "phagocytes" and had confirmed the familiar observation that they played the role of scavengers during physiological and pathological processes.[100] The sixth edition (1889) added nothing to the discussion of phagocytosis but reflected Ziegler's preference for Cohnheim's theory of inflammation. In 1892 Ziegler openly attacked the phagocytic theory. Mechnikov's approach to inflammation was dogmatic and one-sided, and he was so "under the spell of his doctrine of phagocytosis" that he was blind to the negative aspects of inflammation that were so familiar to the clinician. There was no question that "in certain cases phagocytosis may help to destroy foreign bodies," but in others it contributed to the pathological condition. For instance, in cases of leprosy the bacteria multiplied mightily within the white cells and were transported by them throughout the body, spreading the disease. For Ziegler, then, Mechnikov's insistence that phagocytosis was the essence of inflammation, and that inflammation was an organism's defensive response to infection, was hopelessly one-sided.[101]

Darwin and the Defense of the Phagocytic Theory

Beginning in the early 1890s, Mechnikov's comments about Darwin's theory (but not "Darwinism" or "Darwinists") grew increasingly friendly. The striking change in the content and tone of his remarks permits us to speak with but slight exaggeration of two Mechnikovs.

The first was a Russian zoologist writing in the 1860s and 1870s. This Mechnikov was passionately concerned with evolutionary theory per se and wrote constantly about it, drawing on his own work in embryology and zoology. For him Darwin's theory was a powerful but tainted hypothesis, and the social milieu in which he lived encouraged scientific alternatives to its dubious Malthusian aspects.

The second Mechnikov was a pathologist working in France at the turn of the century. He was interested in evolutionary theory primarily for its relationship to phagocytosis. He did not write a single article about evolutionary theory per se, nor did he mention it, except in passing, in his copious letters to his wife. The intellectuals of his adopted home had never welcomed even those aspects of Darwin's theory that he had always found compelling, and an alarming number of them endorsed what Mechnikov considered to be antiscientific alternatives to it.[102]

This second Mechnikov embraced the selection theory and proudly identified his phagocytic theory with it. For him, Darwin had provided the best available response to those pathologists who dismissed the phagocytic theory as teleological and to those philosophers who denied science's capacity to explain the seemingly purposeful qualities of nature. Scattered comments reveal that Mechnikov retained his earlier reservations about Darwin's theory, but these had become quite secondary.

The first Mechnikov did not associate his phagocytic theory with Darwin's ideas. In his initial, confident communication of 1883, he linked it simply with an "evolutionary point of view."[103] Eight years later the second Mechnikov first invoked Darwin's name when responding to his critics in *Lectures on the Comparative Pathology of Inflammation*.

Originating as a series of addresses at the Pasteur Institute in 1891, Mechnikov's *Lectures* reflected his conclusion that resistance to his theory was rooted in medical theorists' lack of a broad evolutionary perspective. To remedy this he proposed a new field, comparative pathology, which was a branch of zoology charged with examining the relations between infectious agents and their hosts. In *Lectures* Mechnikov traced such interspecific conflicts through progressively more complex forms, beginning with the struggle of unicellular organisms against invading bacilli and continuing through simple multicellular forms, metazoans, coelenterates and echinoderms, arthropods and mollusks, and finally amphioxus. Since amphioxus represented "the last survivor of the lower vertebrates," he considered it unnecessary to address more complex organisms of the same order.[104] The remaining chapters of *Lectures* were instead devoted to specific issues such as the nature of leukocytes, the role of blood vessels, and competing theories of inflammation.

This volume revolved around Mechnikov's central dogma that "infection [is] a struggle between two organisms."[105] More specifically, infection was but one in a series of interspecific conflicts that took the form of a host-parasite relationship. He reminded his audience of the prevalence of active interspecific conflict among higher organisms:

> If we examine the organization of an animal or a plant, we find that their most characteristic features are their organs of attack and defense. The carapace of the crayfish, the shell of molluscs and the teeth of the vertebrates, as well as many other organs, are so many means of protection to these animals in their perpetual warfare.[106]

Infection was but warfare by other means. Most *Acinetae*, for example, were equipped with suckers that enabled them to attach themselves to waterborne objects, to draw other infusoria toward them, and to absorb their contents. Other *Acinetae*, however, were free-floating and of such minute size that they were able to penetrate their quarry and lead a parasitic existence within it. Thus, two different members of the same class of organisms could act either as "voracious aggressors or as parasites with the power of producing a definite infection."[107] In the latter instance, interspecific struggle took a more complex form: "The parasite makes its onslaught by secreting toxic or solvent substances, and defends itself by paralysing the digestive and expulsive activity of its host; while the latter exercises a deleterious influence on the aggressor by digesting it and turning it out of the body, and defends itself by the secretions with which it surrounds itself."[108]

The following passage is especially revealing. Mechnikov was defending the phagocytic theory against attacks from all directions and was relying on the selection theory for support, yet he permitted himself a digression that distinguished his approach from that of "orthodox Darwinists":

> Although these phenomena do not come under the heading of the struggle for existence in the strictly Darwinian sense (i.e., competition for the survival of the fittest among individuals of the same species), yet they are all more or less directly connected with the struggle for survival that is always going on between the representatives of the different orders of living beings.[109]

As in earlier years, then, Mechnikov distinguished carefully between the different aspects of the struggle for existence, emphasized interspecific conflict in his own thinking, and associated "strict Darwinians" with a dogmatic insistence on the primacy of overpopulation and intraspecific competition.

Yet this subtle distinction now paled in importance beside the great merit of the selection theory, which allowed Mechnikov to refute his critics. Darwin had explained "the adaptation of means to an end" by demonstrating "that only the characteristics which are advantageous to the organism survive in the struggle for existence, while those that are harmful to the individual are readily eliminated by natural selection."[110] Since the phagocytic theory was based on this insight, it was entirely inaccurate "to attribute a teleological character to the theory that inflammation is a reaction of the organism against injurious agencies."[111] He made this same point in an article of 1895: having "developed entirely on the basis of Darwin's evolutionary theory," Mechnikov's doctrine was innocent of any reliance on "predetermined purposefulness."[112]

The specific use to which Mechnikov put Darwin's theory is also evident from his response to the pathologist Paul Baumgarten. Baumgarten had argued that phagocytes could hardly constitute defensive forces since they often "refuse to act just when the organism is most threatened." Mechnikov responded exactly as had Darwin to similar objections. This lack of perfection was mysterious only to the teleologist; it was entirely understandable to the selectionist. The curative forces of the organism were imperfect and evolving, and hence often failed to repel invaders. Those organisms in which phagocytosis was poorly developed tended to fall prey to parasites, and so "the activity of the phagocytes will increase with every successive generation."[113] Many children died every year because they were "poorly armed for the struggle with microbes." Survivors tended to have more advanced antimicrobial defenses. "This is the process of natural selection, playing itself out daily before our eyes."[114] Precisely because phagocytosis was an imperfect, evolving mechanism, examination of it promised to illuminate "the universal law of natural selection."[115]

Here Mechnikov's reliance on natural selection led him to rely also on intraspecific competition as an explanation of the evolution of phagocytic properties. As he had acknowledged in the 1870s, from a selectionist perspective only this aspect of the struggle for existence could, in the final analysis, generate new species. As in his essays of the 1870s, however, it was not overpopulation but the conflicting needs of different species that generated the struggle for existence and the emergence of new forms.[116]

Mechnikov's assessment of the strengths and weaknesses of Darwin's theory was also apparent in his essay "The Law of Life" (1891). Responding to Tolstoy's critique of natural science in general and Darwin in particular, Mechnikov defended Darwin as a symbol of scientific rationality and as the author of the only available scientific explanation of dynamic harmony in nature. Even so, he disassociated himself from the "rabid Darwinist" Seidlitz (who had reduced the struggle for existence to intraspecific conflict) and referred readers to his "Essay" of 1876, reminding them that here he had "presented in scientific form the entire significance of the Darwinist doctrine, without concealing, it goes without saying, its weaknesses."[117]

Five years later he specifically declined to label himself a Darwinist. In his enthusiastic review of *Charles Darwin and His Theory* (1896), the Russian publicist M. A. Antonovich mentioned the following "outstanding Russian scholars accepting Darwin's theory": Beketov, Timiriazev, and A. O. Kovalevskii.[118] Antonovich, apparently, had asked Mechnikov whether to include him in this list, eliciting the response that Mechnikov, "highly valuing Darwin's theory and recognizing its enormous influence on the development of biology, does not completely accept it and does not consider himself a Darwinist on the basis that the new theory has not conclusively resolved all the questions pertaining to it and that there are in it points requiring further investigation and factual support."[119] A disappointed Antonovich termed Mechnikov's comments "very prudent but also extremely banal," since one could adopt such a view of any theory. But Mechnikov was not simply splitting hairs. It was one thing to see natural selection as an important contribution (as he had even in the 1870s), and even to view it as a necessary component of his phagocytic theory, and quite another to associate himself with "Darwinists" who sought to explain evolution almost exclusively in terms of overpopulation, intraspecific competition, and natural selection.

In his discussions of aging, as in his work on inflammation, Mechnikov sometimes digressed to distinguish between the different aspects of the struggle for existence. He wrote in *The Nature of Man*:

> A conflict takes place in old age between the higher elements and the simpler or primitive elements of the organism, and the conflict ends in the victory of the latter. This victory is signalised by a weakening of the intellect, by digestive troubles, and by lack of sufficient oxygen in the blood. The word conflict is not used metaphorically in this case. It is a veritable battle that rages in the innermost recesses of our beings.[120]

Mechnikov's final published comments on Darwin and Darwinism appeared in three short articles (two for the Russian audience and one for the French) of 1909 and 1910. Each was written for a celebration of Darwin's achievement, and in each he praised Darwin highly.[121] Darwin had discovered that natural selection regulated harmony in the natural world and so had struck a blow against metaphysics.[122] While "not a few gaps were found" in Darwin's theory, it had become part of the "flesh and blood of positive science."[123] Mechnikov also gave Darwin credit for his own Nobel Prize-winning work: "Based on the principles of Darwin's theory, it was possible to convincingly demonstrate that inflammation, one of the fundamental characteristics of an entire series of pathological processes, represents a defensive act of the organism against pathogens."[124] Medical science was returning Darwin's favor by demonstrating that the evolution of illnesses was governed by "the great laws discovered by Darwin."[125] For instance, new strains of relapsing fever seemed to result from certain microbes that "due to their individual characteristics, avoided the action of the curative forces of the organism" and produced more highly resistant offspring. This, Mechnikov suggested, might provide an example of evolutionary change "generated by the difference between individuals of one and the same species." While this statement was certainly supportive of Darwin's theory, it also implied that such examples were still necessary in order to substantiate it.[126]

Even in these celebratory articles, Mechnikov criticized "orthodox Darwinists" for their failure to amend Darwin's views. He noted proudly that a Russian botanist, Korzhinskii, had been among the first to revise Darwin's ideas in light of new research on variability and heredity, and he commented favorably on de Vries's mutation theory.[127] Here again he made evident his continued skepticism regarding Darwin's approach to the struggle for existence. He recalled his attempt in the early 1870s to test Darwin's explanation of the evolution of wingless insects on the island of Madeira. Darwin had suggested that short-winged insects were less likely to be caught in the wind and drowned offshore, and so were more successful in the struggle for existence than were their longer-winged fellows. Mechnikov recalled that he had compared the wing length of drowned insects with those on dry land and had found no difference, confirming his doubts about Darwin's argument.[128]

In his article for a French audience, Mechnikov dismissed "orthodox Darwinists" as hopelessly outdated but concluded that "neo-Darwinism has a serious basis." He informed his Russian audience that "if it is impossible to consign orthodox Darwinism to the archive, then, obviously, many of Darwin's propositions should be subject to a new revision."[129]

The Death of Il'ia Il'ich

In "The Death of Ivan Il'ich" Tolstoy used the final days of Mechnikov's older brother to illustrate the empty mendacity of conventional life and the pompous impotence of medical science. Mechnikov was present at his brother's deathbed and commented briefly on Ivan's death in his *Prolongation of Life*, but his own decline and demise offered a much better opportunity to fashion a well-choreographed counterpoint to Tolstoy's story.

Several chapters of Ol'ga Mechnikov's biography, *Life of Elie Metchnikoff*, could indeed be appropriately entitled "The Death of Il'ia Il'ich."[130] These pages illustrate the correctness of Mechnikov's theory of aging and the power of scientific rationalism. "At this supreme moment of Elie Metchnikoff's existence, everything was full of significance," Ol'ga wrote, "for everything converged to emphasize the powerful unity and the ascending and continuous progress of his ideas. His attitude in the face of illness and death was a teaching, a support, and an example."[131]

Tolstoy's Ivan Il'ich had rued the emptiness of his life and feared his premature death, the horror of which was magnified by the evasions and false assurances of polite company. The Mechnikovs' Il'ia Il'ich reflected contentedly on a life well spent, observed the progress of his disease with cool, scientific objectivity, and took comfort in the harmonious waning of his life and his life instinct. As his condition worsened he made the following observations:

> I have taken no raw food for eighteen years and I introduce as many lactic bacilli as possible into my intestines. But it is but a first step; in spite of all, I am being poisoned by the bacteria of butyric fermentation. However, I must be satisfied. I have, so to speak, accomplished the programme of a "reduced orthobiosis."[132]

In his final moments Tolstoy's Ivan Il'ich had achieved transcendental clarity and welcomed death. In his, Il'ia Il'ich delivered some parting instructions regarding the postmortem: "Look at the intestines carefully, for I think there is something there now."[133]

In the course of their amiable discussions at Yasnaya Polyana in 1909, Tolstoy and Mechnikov had reportedly agreed that they were both "pursuing the same goal of human perfection and happiness, but going along such different roads."[134] Mechnikov's had taken him from beet weevils to phagocytes, from a critique of Darwin's Malthusianism to a reliance on his discovery of natural selection, and from a despairing recognition of natural disharmony to a triumphant recognition of science's ability to correct it. Yet over these many years one thing had remained unchanged: his critical attention to the dynamics of the struggle for existence and his preoccupation with the role of interspecific conflict. "Let those who will have preserved the combative instinct," he wrote during World War I, "direct it towards a struggle, not against human beings, but against the innumerable microbes, visible or invisible, which threaten us on all sides and prevent us from accomplishing the normal and complete cycle of our existence."[135]

CHAPTER 6

Kessler and Russia's Mutual Aid Tradition

And another interesting thing: Kessler, Severtsov, Menzbir, Brandt—4 *great* Russian zoologists, and a 5th lesser one, Poliakov, and finally myself, a simple traveller, stand against the Darwinist exaggeration of struggle within a species. *We see a great deal of mutual aid*, where Darwin and Wallace see *only struggle*.[1] (P. A. Kropotkin, 1909)

The goal of science is to humanize man and to direct his powers to useful activity.[2] (K. F. Kessler)

Western readers will probably associate the theory of mutual aid with Petr Kropotkin and his anarchist political philosophy. Kropotkin's views, however, were but one expression of a broad current in Russian evolutionary thought that predated, indeed encouraged, his work on this subject and was by no means confined to leftist thinkers. For Russians the seminal mutual aid theorist was a well-established naturalist with centrist political views, K. F. Kessler, whose 1879 speech "On the Law of Mutual Aid" transformed a widespread sentiment into a coherent intellectual tradition.

We should distinguish between "weak" and "strong" variants of this mutual aid tradition. N. A. Severtsov and M. A. Menzbir exemplified the former: they wrote a great deal about the importance of mutual aid in nature but saw no necessary tension between this phenomenon and Darwin's theory.[3] I also include here naturalists such as I. S. Poliakov who emphasized cooperative relations among organisms but whose work lacked explicit theoretical content.[4] The publications of such authors would almost certainly have appealed to Darwin, who, after all, himself wrote quite a bit about cooperation in nature in *The Descent of Man*.

Proponents of the "strong" variant, however, went far beyond Darwin in their acceptance of four basic tenets: (1) The central aspect of the struggle for existence is the organism's struggle with physical circumstances (or, less frequently, with members of other species); (2) organisms join forces to wage this struggle more effectively, and such mutual aid is favored by natural selection; (3) since cooperation, not competition, dominates intraspecific relations, Darwin's Malthusian characterization of those relations is false; and (4) mutual aid so vitiates intraspecific conflict that this cannot be the chief cause of the divergence of characters and the origin of new species. Implicit in this line of argument was the juxtaposition of cooperation and conflict (both passive competition and direct struggle). These were treated almost as if they were opposed physical forces; to the extent that one was present, the other was necessarily absent.

Among Russian biologists subscribing to this "strong" position at one time or another were Beketov, Mechnikov, physiologist-psychiatrist V. M. Bekhterev, zoologist M. N. Bogdanov, morphologist A. F. Brandt, soil scientist V. V. Dokuchaev, embryologist N. D. Nozhin, zoologist K. F. Kessler, geographer-geologist P. A. Kropotkin, and hygienist I. S. Skvortsev. This view was also espoused by numerous populist theoreticians, including P. L. Lavrov and N. K. Mikhailovskii, by the encyclopedist M. M. Filippov, and by the theologian M. Glubokovskii.[5]

The political views and scientific experiences of the scientists identified above varied greatly. As a group they inclined somewhat to the political Left (Kessler and Brandt were important exceptions) but probably no more so than the Russian intelligentsia as a whole. They shared a cooperative ethos that was expressed both in explicitly prescriptive terms, drawing on populist, conservative, and religious ideologies, and in simple descriptive terms: the physical conditions of life were harsh, forcing organisms to draw upon all available resources, including each other. To this was often joined a belief that natural law had prescriptive implications for human society and a perception that those of Darwin's struggle for existence were fallacious and sordid. Roughly one-half of the mutual aid theorists I have identified were field naturalists, but Bekhterev, Brandt, Mechnikov, Nozhin, and Skvortsov were not. The mutual aid tradition was strongest in St. Petersburg. Beketov, Bogdanov, and Brandt all taught at St. Petersburg University and acknowledged Kessler's influence on their views. Bekhterev, Dokuchaev, and Kropotkin also lived in the capital at one time in their careers.

In the 1860s and 1870s scientists and lay intellectuals often asserted that cooperation in nature belied Darwin's emphasis on competition. We have already seen the commonsensical appeal of this view in eighteen-year-old Il'ia Mechnikov's initial reaction to the *Origin*. "As everyone knows," he wrote in 1863, "common dangers and obstacles do not stimulate struggle between the individuals subject to these misfortunes, but rather cause them to unite in one society, to resist more successfully the obstacles facing them through joint action."[6] Three years later, in a widely discussed article, Mechnikov's friend Nozhin also observed that under normal conditions "identical organisms do not struggle for existence among themselves but tend, on the contrary, to combine with one another, to unite their forces and interests."[7]

Such comments were scattered among popular and scientific articles about Darwin's theory until 1879, when they received the imprimatur of one of Russia's leading naturalists.

K. F. Kessler

Karl Fedorovich Kessler (1815–1881) was born in November 1815 in Königsberg, East Prussia. Seven years later his father accepted a position as chief forester for the grenadiers encampments in Novgorod province, and the family moved to Russia.

After graduating from a St. Petersburg gymnasium, Kessler entered the physicomathematical faculty of St. Petersburg University in 1834. His first love was pure mathematics, and his senior essay on that subject won him a gold medal. He also studied zoology with Professor S. S. Kutorga and with F. F. Brandt, founder and director of the Zoological Museum of the Academy of Sciences. In 1837 he traveled to Finland on his first zoological expedition together with his friend N. I. Zheleznov, later a professor of botany at Moscow University and the first director of the Petrovsk Academy of Agriculture and Forestry.

After graduation in 1838 he remained in St. Petersburg, where he taught mathematics and physics at a gymnasium (Beketov was one of his pupils). There Kessler participated in a literary circle whose members circulated their manuscripts anonymously in order to encourage candid reactions. He contributed several poems praising Napoleon, and the response convinced him to abandon this avocation.

His interest soon turned decisively toward zoology. Continuing his work with Kutorga and Brandt, he defended a monograph on birds in 1840 for his master's degree. Two years later he received a doctorate for his thesis on the skeleton of the woodpecker. Together with Zheleznov and zoologist A. F. Middendorf, he spent his summers hunting and collecting.

In 1842 Middendorf left the University of St. Vladimir (later Kiev University) to explore Siberia, and Brandt prevailed upon the Ministry of Education to hire Kessler in his place. He was promoted to full professor in 1845, became dean of the physicomathematical faculty in 1854, and remained at Kiev University until 1861.

The university had inherited its zoological collection from a failed lycée and the Vilensk Medical-Surgical Academy, and Kessler appraised it with some dismay. The great majority of the specimens were mollusks and insects. Intent on building the collection and attracted by the beauty of the Kiev area, Kessler spent his free days from early spring until late fall in pursuit of local wildlife. Botanist R. E. Trautfetter, naturalist A. S. Rogovich, and a local obstetrician skilled at the hunt accompanied him on his frequent trips.[8] During his years at Kiev University, Kessler traveled by horseback and boat on expeditions throughout the Kiev, Poltava and Chernigov regions. He also accompanied a military expedition to central Asia in 1848 and studied the fish of Bessarabia in 1856 and those of the Black Sea in 1858.

These expeditions sparked an interest in the periodic phenomena of animal life, particularly the migration of birds. Yet Kessler identified two obstacles to comprehensive work on this subject. First, no single person could monitor migrating birds, which led Kessler to establish a network of observers (and to appreciate the benefits of collaboration). Second, there was no Russian text of systematic descriptions. Kessler's published work in this period was intended to fill this gap and reflected his increasing interest in zoogeographical issues. Most important were his five volumes, published in the years 1851–1856, on the mammals, fish, and birds of the Kiev region.[9]

In 1845 he spent four months in western Europe, traveling to Vienna, Munich,

Zurich, Paris, Leyden, Amsterdam, and Leipzig. He met Ernst Haeckel, with whom he later corresponded about ichthyology, and discussed zoology with Auguste de Saint Hilaire.

Kutorga had trained Kessler according to Cuvierian orthodoxy, and Kessler's early publications reflected his belief in the fixity of species. By the late 1840s, however, he had begun to waver, as was evident in his article "On the Origin of Domestic Animals" (1847). Here he contended that domestic forms had been modified from their wild forebears by the direct action of the environment and by their "moral education" by man—the results of which were passed to progeny by the inheritance of acquired characteristics.[10] During a trip to Moscow in 1851, Kessler met evolutionist K. F. Rul'e. They corresponded in subsequent years, and Kessler collaborated on Rul'e's journal *Vestnik estestvennykh nauk* (*Herald of the Natural Sciences*). By the mid-1850s Kessler was firmly in the transmutationist camp.

While at Kiev University Kessler began to lobby for regular national meetings of Russian naturalists, which, he argued, were especially important given the country's great size. His attendance at western European scientific conferences had impressed upon him their usefulness in stimulating competition, facilitating collaboration on projects beyond the efforts of a single investigator, and resolving controversial issues. In a letter of 1856 to A. S. Norov, an official in the Ministry of Education, he observed that such gatherings were especially necessary in Russia, whose enormous expanse created great difficulties for naturalists. It was not uncommon, he wrote, for a Russian zoologist or botanist to work for a decade without encountering a single other member of his discipline. Perhaps thinking of his own interest in migration, he added: "Such isolation of scholars has the most harmful influence on the strength and success of their scientific activity, placing insurmountable obstacles before all scholarly endeavors requiring the joint activities of many people."[11] Kessler proposed biannual conferences of naturalists and physicians, and suggested that when Russia's railroad system was sufficiently developed, these might become yearly events. The Ministry of Education did not act on his proposal, but Kessler did manage to convene a successful conference of Kiev's naturalists and teachers of natural science.

When Kutorga died in 1861, Kessler replaced him as professor of zoology at St. Petersburg University. The authorities closed the university that year in response to student demonstrations, so Kessler postponed his arrival until 1862. He remained there until his death in 1881. A well-liked man who also enjoyed the confidence of the authorities, he served as rector of the university from 1867 until 1873, when he resigned, claiming poor health. In his efforts to build the department of zoology, he brought Mechnikov to the university as dozent for the academic year 1868–1869 and also recruited A. F. Brandt, N. P. Vagner, and M. N. Bogdanov in the 1870s. Bogdanov became his friend and closest co-worker.

His position at St. Petersburg University enabled Kessler to become a leading organizational figure in Russian science during a period of vigorous institutional expansion. The prestige of the natural sciences swelled in the late 1850s and 1860s, boosted by the scientism of the radical intelligentsia and by the perception among ruling circles, especially after Russia's defeat in the Crimean War, that a productive scientific community was important to the power and prosperity of the state. Minister of Education D. A. Tolstoy endorsed Kessler's earlier proposal for regular gatherings of Russian scientists, and in 1867 Kessler presided over the first Congress of Russian Naturalists and Physicians. The six hundred participants decided to found a society of naturalists at each university, which quickly became important centers for planning, subsidizing, and publishing scientific work. In 1868 Kessler was unanimously elected president of the St. Petersburg Society of Naturalists, a position he held until his death. His old acquaintance Beketov, now professor of botany at St. Petersburg University, served as secretary of the society and editor of its journal.

On his arrival in St. Petersburg, Kessler had intended to continue his previously eclectic research in zoology, and requested permission to collect zoological specimens freely in the St. Petersburg area. Most of the land in question, however, was reserved as the tsar's hunting preserve, and so this request was refused. He concentrated, therefore, on fish, quickly emerging as Russia's leading ichthyologist and the beneficiary of specimens collected by other investigators. In 1871–1872 he wrote an article, entitled "The Ichthyological Fauna of Turkestan," based entirely on material gathered by N. A. Severtsov in central Asia. He also acquired rich collections from the trips of Bogdanov to Khiva in 1873, A. P. Fedchenko to Amu Darya in 1874, N. M. Przheval'skii to northwestern China in 1876, and a contingent sent by the St. Petersburg Society of Naturalists to the Aralo-Caspian region in 1874–1876.

Ichthyology was a specialty with great significance for the tsarist economy. Like von Baer and Danilevskii, Kessler sought to use scientific knowledge to improve the harvest of fish and caviar. In what were apparently his only lectures to a nonacademic audience, he delivered eight addresses to the Imperial Free Economic Society in 1863 on the classification, reproduction, and life of fish.

Kessler relied on nonacademics for a great deal of his information. In his volume on the fish of the Aralo-Caspian region (1877), he conceded that "almost all information on the life of fish is taken from fishermen."[12] Although their testimony was not always reliable, fishermen knew when and where fish collected in large numbers in order to migrate (and so were most efficiently caught); Kessler found this information invaluable.

His expeditions along the Volga, and in the Crimea and the Caucasus left Kessler frustrated with the limitations of purely morphological investigations.[13] By the early 1870s he referred constantly to the need for a broad zoogeographical approach in order to resolve "various interesting zoological and physicogeographical questions."[14] This growing conviction that zoogeography held the key to "the resolution of the great question of the origin and formation of species" exercised a profound influence, not only on his own work, but also on the direction of zoological and botanical research in St. Petersburg.[15] Under Kessler's leadership the St. Petersburg Society of Naturalists sponsored expeditions throughout the country in order to examine "the dependence of the flora and fauna upon the geological-physical conditions of the land, on the one hand, and mutually upon one another, on the other." This, Kessler suggested, would illuminate "the laws governing the natural world."[16]

His own comments about biological theory remained quite circumspect until he delivered his address "On the Law of Mutual Aid" to the St. Petersburg Society of

Naturalists in December 1879. This speech excited general enthusiasm, but Kessler died two years later without having elaborated his ideas.

One year after Kessler's death, Bogdanov honored him with a biography. For Bogdanov, Kessler's discovery of the law of mutual aid was the logical culmination of his work, not only as a zoologist, but also as an organizer of Russian science. "Before presenting this law in scientific form did he not put it into practice, into life! Did this law not guide him in his social activity, in his consistent effort to unite Russia's scientific family!"[17]

Ichthyology and the Law of Mutual Aid

Kessler's scattered comments on evolution in the years 1863–1877 suggest that he welcomed Darwin's theory but, at least by the early 1870s, thought natural selection of only secondary importance for evolution. They also reveal that his theory of mutual aid was not based on a benign vision of natural relations in general or of intraspecific relations in particular.

In his lectures to the Imperial Free Economic Society in 1863, Kessler observed that the life of fish, like that of all animals (including man), was governed by the inborn drives to obtain nourishment and to reproduce. In their search for food

> fish conduct an incessant, extremely harsh war among themselves, the strong pursue the weak, the large devour the small, sparing not even their own offspring; a twenty-pound pike swallows a five-pound pike. Nowhere does there occur such a constant and intense struggle for existence as in the silent kingdom of fish under the bright surface of the water.[18]

Kessler mentioned Darwin only in passing, but was clearly familiar with the *Origin*.[19]

He first addressed the selection theory in an 1865 review of literature on domesticated animals. Defending Darwin against a critic, Kessler sketched the selection theory briefly and sympathetically: species changed gradually through the natural selection of favored individuals, a process generated by overpopulation and the struggle for existence. He agreed with Darwin that it was "understandable" that this struggle would be most intense "among those individuals who are placed in identical living conditions and require identical food, that is, among individuals that are most alike."[20]

Although he frequently alluded to his interest in the factors of evolution, and presented Darwin's and Lamarck's theories to his students at the university, Kessler first advanced his own tentative ideas in a massive tome on the fish of the Aralo-Caspian region (1877). Here he made no mention of Darwin, natural selection, or the struggle for existence. He suggested that unknown "internal causes," which were probably regulated by environmental influences, "condition differences between individuals." These differences were then "accumulated by the force of heredity and consolidated by external causes, by the external conditions of the life of the animal. In this manner, varieties of existing species are initially formed, and new species are then produced from these varieties."[21] The direct action of the environment, then, combined with geographical isolation to produce new forms.

Acknowledging his debt to Moritz Wagner, Kessler illustrated the evolutionary role of isolation with several examples gathered during his expeditions. Fish that traveled singly or in small groups sometimes became isolated in different mountain waterways with varied conditions of existence. The influence of these varied environments had produced various species. Kessler had discovered seven different species of the family Benthphilus in one water basin; each inhabited a different depth with different life conditions. This demonstrated "the powerful influence of the environment upon animals."[22]

His emphasis on isolation and the direct influence of the environment, rather than on natural selection, did not reflect a newfound belief that relations among fish were benign and peaceful. He repeated his earlier assertion that "in no other class of the animal kingdom does there proceed . . . such a terrible mutual extermination of some species by others, and even of the weakest individuals of a species by the stronger individuals of the same type, as among fish."[23]

The "life relations" among fish, however, were fluid and subject to two distinct influences: while the drive for food generated a harsh, individualistic struggle, the drive to reproduce often led fish to live peacefully together. For example, spawning fish often traveled from salty to fresh or from standing to flowing water. Those, like sturgeon, that traveled great distances usually did so in large schools.[24] Kessler observed that the differing degrees of sociability among the fish of the Aralo-Caspian region exercised a powerful influence on "the formation of [individual] differences and new species":

> Among fish that unite into large schools while spawning and undertake significant travels from one place to another . . . sharply marked differences are much rarer than among fish that hold more to themselves or travel in small colonies while spawning, and which do not undertake long migrations. This, in my opinion, explains the origin of the many, often very closely related species of fish that populate the rivers and lakes of the mountainous lands of our region. In the mountain waters, with their varied external conditions, [individual] differences are first established and then, with the more or less complete isolation of these forms, new species appear.[25]

Kessler had always introduced his discussion of life relations among fish by calling attention to the basic drives to eat and reproduce. With his 1877 volume, however, he began to discuss the influence of each drive on "life relations" and the impact of varied "life relations" on evolution.

During his travels he also had ample opportunity to observe peasant life. Describing it in 1878, he distinguished between those conditions common to all peasants and those arising from competition among them. The peasantry, he observed, suffered chiefly from an unforgiving natural environment and from general impoverishment. Yet peasants also struggled among themselves: those with sufficient capital to hire labor exploited their fellows unmercifully.[26]

The relevance of the struggle for existence to human relations was clearly on Kessler's mind when, one year later, at age sixty-four, he addressed evolutionary theory at length for the first and last time. He began his address "On the Law of Mutual Aid" by criticizing those who invoked "the cruel, so-called law of the struggle for existence" to resolve social and moral issues.[27]

Kessler reminded his audience that Darwin's struggle for existence included both the "open warfare" of predator-prey relations and the indirect "hidden struggle" that often transpired within a family or species. He agreed that overpopulation often generated intraspecific competition and conceded, as he had since 1863, that "the struggle for food between individuals of one species, or between close species of one family, is often the cruelest, most merciless of all."[28]

Yet Darwin exaggerated the prevalence of intraspecific conflict and "too one-sidedly relies on the struggle for existence." Zoology and the sciences of man had both ignored the existence of "another law, which one could term the law of mutual aid, which, at least in relation to animals, is if anything more important than the law of the struggle for existence."[29]

Just as the struggle for existence originated in the organism's need to obtain food, so did mutual aid result from the drive to reproduce. "The simple inclination of animals to reproduce is transformed into an attraction between some animals and others, and, as a result of this mutual attraction, there appears a certain sociability among individuals of the same species." Sexual attraction and reproduction also led adults to care for the young. Thus, "separate individuals cease to be concerned only with feeding and preserving themselves, but begin to aid other individuals."[30] Mutual aid within a species vitiated conflict among its members and facilitated the battle against other species.

Kessler illustrated this point with examples, drawn largely from his own travels, of mutual aid among bees, ants, beetles, spiders, fish, reptiles, birds, and mammals. The importance of "family and social life" among birds, for instance, was "stunning." He recalled one instance in which a duck and its offspring were threatened by hunters. The duck fled, seemingly abandoning its defenseless young, but soon returned with a drake. While the drake distracted the hunters, the duck removed its offspring to safety.[31] Even birds of different species sometimes aided one another, as in migratory flights. While traveling in the Crimea Kessler had often seen colonies of different species playing happily together, enjoying the material and spiritual advantages afforded by mutual aid. "Some like to entertain one another with song, others enjoy various flying competitions, still others find satisfaction in dance and in bloodless duels before a crowd of their fellows."[32] Different organisms had attained various degrees of mutualism. For example, family life among mammals was less developed than among birds. Mammalian "marital unions are less frequently concluded for a lifetime, and one sees polygamy or polyandry more frequently" among them.[33] Domesticated animals were among the least sociable of creatures.

Kessler's analysis of the relationship between mutual aid and the struggle for existence was inconsistent. In some instances he portrayed them as polar opposites, arguing that mutual aid was more important than the struggle for existence. In others he spoke of them as complementary, contending, for instance, that mammals often experience "a need for mutual aid in the struggle for existence."[34]

For the most part Kessler employed mutual aid simply as an ecological concept—a description of natural relations—yet he did sketch its evolutionary significance in the following terms: individuals within a species were exposed to the influence of differing environments, which produced differences among them. This conferred a competitive advantage upon some of them. These favored variants could only increase their number and acquire evolutionary significance, however, by reproducing. This, in turn, required mutual aid.[35] Therefore, a species in which cooperative relations were more highly developed had a greater chance to "proceed further in its development and improvement, including intellectually."[36] Mutual aid, then, increased the likelihood that the new forms created by the direct action of the environment would survive.

"I do not reject the struggle for existence," Kessler insisted, "but only affirm that the progressive development both of the entire animal kingdom and, especially, of mankind is not facilitated by mutual struggle so much as by mutual aid."[37] He summarized his argument as follows:

> All organic beings have two fundamental requirements: nutrition and reproduction. The need for nutriment leads them to the struggle for existence and mutual extermination, while the need to reproduce leads them to approach and support one another. But for the development of the organic world, the transformation of one form into another, the uniting of individuals within the same species is if anything more important.[38]

This was especially true of man. "The law of mutual aid has played an incomparably greater role in the history of his successes than has the law of the struggle for existence." Mutual aid underlay humanity's material and moral progress. Its increasing influence was evident in the friendships formed between former enemies after a war and in the attempts of stronger tribes to assist weaker ones. Clearly, sometime in the future "all people will consider themselves brothers," and bloody confrontations will be replaced by peaceful competition in industry and science.[39]

The response to Kessler's speech was, by all accounts, enthusiastic. Brandt later recalled "the noisy approval of the large audience."[40] Severtsov applauded Kessler's insight and added a supportive observation about mutual aid among falcons; another Russian naturalist later claimed that "the correctness of [Kessler's] views struck most of the Russian zoologists present."[41]

When Kessler died in 1881, an anonymous "student of the 1860s" composed a poem in his honor. In a century of "money grubbing and alienation," the student wrote, Kessler had "upheld the lamp of love." By his insistence that "unity and aid to those near," rather than "struggle and extermination," was a basic principle of nature, he had summoned humankind to an approaching era of love and brotherhood.[42] Although this commemorative effort did not fully communicate the logic of Kessler's argument, it certainly captured an important element of its broad appeal for Russians.

Mutual Aid after Kessler

After Kessler's speech the "strong" mutual aid argument became a common feature of Russian evolutionary thought. Most of its proponents praised Darwin highly despite their criticism of his distorted view of intraspecific relations. They were less charitable toward "Darwinists," who had raised Darwin's error to a dogma that served as the "slogan and war cry of the day."[43]

For many Russians the theory of mutual aid simply expressed a gut sentiment about nature and morality. Consider, for example, soil scientist V. V. Dokuchaev's comment:

> The great Darwin, to whom contemporary science is indebted for perhaps 9/10 of its present scope, thought that the world is governed by the Old Testament law: an eye for an eye, a tooth for a tooth. This is a big mistake, a great confusion. One should not blame Darwin for this error, and one cannot attribute it to a lack of talent.... But Darwin, thank God, turns out to have been incorrect. Alongside the cruel, strict Old Testament law of constant struggle we now see clearly the law of cooperation, of love. And nowhere is this law so sharply and obviously manifest as in the doctrine about soil zones, where we observed the closest mutual interaction and complete cooperation of the organic and inorganic worlds.[44]

For Dokuchaev, Darwin's mistake was one of emphasis. Struggle did, in fact, characterize life relations in some tightly constricted geographical areas. One need only glance at Russia's broad expanses, however, to see that "over stretches of thousands of versts of black soil," organisms "accommodate and complement one another."[45]

The scientific status of mutual aid was greatly enhanced by Beketov's endorsement in his influential *Textbook of Botany* (1882).[46] Theologians also took heart. Writing for the popular religious journal *Vera i razum* (*Faith and Reason*) in 1892, M. Glubokovskii praised a lecture delivered by Kharkov University's professor of hygiene, I. S. Skvortsev. Skvortsev had demonstrated convincingly that "every step in the development of life, every new form of life, is the consequence, not of struggle, but of mutual interaction, mutual aid."[47]

A closer examination of four mutual aid theorists serves to illustrate the origins, logic, and appeal of this scientific tradition.

A. F. Brandt

Alexander Fedorovich Brandt (1844-?) was a philosophical idealist, a political conservative, and a scholar of the museum and laboratory. The son of zoologist F. F. Brandt, he studied neuro-physiology in the 1860s at the St. Petersburg Military-Medical Academy with I. M. Sechenov (1829–1905). Sechenov was at the time a hero to radical youth for his essay *Reflexes of the Brain* (1863), in which he explained volitional acts as the result of reflex reactions. Brandt subsequently adopted an entirely different philosophical perspective, as expressed in the title of his book *From Materialism to Spiritualism*.[48] After working with Sechenov he studied histology with F. V. Ovsiannikov and spent a summer in Jena with Ernst Haeckel and Karl Gegenbauer. He received his doctorate of medicine in 1867.

In 1871 he succeeded his father as director of the Zoological Museum of the Academy of Sciences and lectured in Kessler's Department of Zoology at St. Petersburg University. He received his doctorate in zoology in 1876 and was appointed professor at Kharkov Veterinary Institute in 1881 and at Kharkov University in 1887. His scientific work covered a wide range of topics, including anthropology, comparative anatomy, and medical zoology.[49]

Brandt wrote regularly for *Niva*, a weekly magazine that provided its middle-class audience with less weighty fare than such "thick journals" as the *European*

Herald and *Fatherland Notes*. The struggle for existence often figured prominently in his articles on science and medicine. For him this expression referred to the complex of relationships that maintained balance in the natural world.

In "The Numerical Balance among Animals in the Struggle for Existence" (1879), Brandt took issue with Malthus's and Darwin's views of population dynamics, arguing that fertility was constantly influenced by "the general conditions with which all animals without exception conduct a struggle for existence."[50] The rate of reproduction thus provided "a truly calculated counterweight" to those conditions.[51] He illustrated this with discussions of infusoria, parasites, and various insects, and commented that the same principle held true for man, who was engaged in a struggle for existence in which "he is far from always a victor."[52]

As an evolutionist Brandt conceded that the numerical balance was occasionally upset by the emergence of new species or by a change in environmental conditions. The appearance of lions and tigers, for example, had certainly diminished the number of antelope. A new balance was rapidly achieved, however, and so "the progressive development of animal life on our planet . . . does not contradict the law of the numerical status quo in the animal world."[53]

In 1882 an agricultural congress was convened in Kharkov to consider measures against pest infestations, and Brandt drew on his understanding of the struggle for existence to explain that problem to his readers. "The entire totality of organisms," he explained, "constitutes a strict, harmonious whole."[54] The struggle for existence between wolves and deer, and between insects and plants, maintained a situation in which all could survive in the same forest. Human agriculture violated this natural balance. By sowing wide areas with cultured plants and displacing hordes of wild animals with a few domesticated ones, humans imposed an unnatural uniformity on cultivated lands. Natural checks to the multiplication of mice, gophers, and other agricultural pests were thus eliminated, so hordes of them competed with humans for the harvests that now provided the only food in the region. The only permanent solution was either to restore the previous natural balance or, where this was impossible, to establish a new, artificial one. Brandt suggested that entomologists and other naturalists familiar with the dynamics of the struggle for existence might introduce carefully selected plants and animals into cultivated areas in order to "bring agriculture into accordance with the demands of nature itself."[55]

For Brandt a stratified, harmonious social system also reflected the natural order. Naturalists knew the plant kingdom provided food both for its own consumption and for that of animals. Similarly, the working class accumulated wealth both for itself and for "nonproducing consumers."[56] While a traditional Russian folk saying emphasized the role of the producers ("without the ploughman there would be no velvet ribbons") the opposite was equally and increasingly true. With the gradual exhaustion of the soil and mineral wealth, the nonproducers would become increasingly important to the management of human industry. Only the division of labor between the two groups made it possible for humans to wage a successful struggle for existence.[57] This struggle was waged collectively by a harmonious and cohesive community, one in which individuals were "equal" in the sense that each was necessary to the success of the group.

Brandt's most complete exposition on the struggle for existence was "Symbiosis and Mutual Aid" (1896). Here he observed that Darwin's concept had become the rallying cry of fin de siècle dilettantes who urged upon mankind "the struggle for existence, the war of all against all, the philosophical system of Friedrich Nietzsche, and the right of the fist in human society."

> It is the zoologist's responsibility to demonstrate that the significance of the struggle for existence in the animal world is exaggerated and to present, in opposition to it, the principle of mutual aid. This principle is also prevalent [in nature] and because of it, not struggle, many animals have come together in families, communities and their own type of government, resembling human ones.[58]

Overpopulation sometimes generated conflict, but it also gave rise to two distinct forms of cooperation. First, purely external factors led such lower organisms as the hydra, the army worm, and the beet weevil to unite among themselves. When food was plentiful beet weevils "conduct their destructive work side by side, and in a friendly manner." Yet this was not yet "communal work" since each individual went about its business "without concern for its comrades." When the supply of readily available grain was exhausted, however, relations among beet weevils were fundamentally transformed: they departed together in search of richer pastures. This migration en masse testified to the awakening of "a social instinct, a sense of solidarity."[59]

The beet weevil thus manifested elements of a second, more developed form of cooperation. This was not simply a reaction to external conditions but was "voluntary" and featured a consistent division of labor and mutual aid. Such cooperation was striking among ants, wasps, and bees, and arose primarily for defense of the young and acquisition of food. The ants of a single anthill usually cooperated, sometimes in a fight to the death against outsiders. So pronounced was the division of labor among them that some were physically incapable of chewing their own food and living outside of the community.[60]

The degree and nature of cooperation among other animals varied widely. Amphibians and reptiles were rarely sociable. Fish were drawn together by sexual attraction and by the rigors of their travels across long distances to spawn. Social life among birds was only periodic. They sometimes cooperated in the search for scarce food, but this tentative communality disappeared when they were sated. The perils of migration led even birds of different species to cooperate. Beavers cooperated during construction projects; Brandt cited zoologist N. A. Kholodkovskii's report that he had seen four of them snap off a tree branch by jumping upon it in coordinated fashion.[61]

For Brandt the distortion of biological theory owed less to Darwin than to "Darwinists." However one-sided Darwin's approach to the struggle for existence, in his hands this concept had constituted a profound contribution to science with implications for "the most varied spheres of investigation, including economics, in which it first arose."[62] Darwin had appreciated the importance of mutual aid but had been too preoccupied with other subjects to explore fully its significance. His successors, however, ignored cooperation entirely and built "an entire cult" around individualistic struggle.[63]

Brandt applauded Kessler's correction to this tendentiousness. His discovery of the law of mutual aid was science's proper response to the exaltation of conflict. "Social animals struggle for existence with external enemies and various adversities, and do so more successfully than do animals who maintain antagonistic relations with their fellows."[64] From which group, Brandt asked rhetorically, should mankind take its example? As a familiar folk saying had it, "Every vegetable has its day"—and competition's day was passing. The abolition of slavery presaged the end of all human conflict, the disappearance of the counterproductive struggles of "class against class, party against party, worker against employer, capitalist against proletariat, buyer against seller." Science and industry would eventually eliminate poverty, and egoism would yield entirely to altruism.[65] Not competition but unity was "the only rational, the only possible slogan of cultured humanity."[66]

M. M. Filippov

Mikhail Mikhailovich Filippov (1858–1903) was a populist intellectual with an encyclopedic range of scholarly interests. A graduate of the physicomathematical faculty of Novorossiisk University and the juridical faculty of St. Petersburg University, he also studied with the chemist Marcelin Berthelot in Paris and received a doctorate in natural philosophy from Heidelburg. A close friend of D. I. Mendeleyev, Filippov translated his *Foundations of Chemistry* into French and also translated the sixth edition of the *Origin* into Russian. He wrote numerous books and articles on mathematics, natural science, philosophy, political economy, history, and literature. These included essays on the sun, evolutionary theory, human races, and the sociology of Henry George, as well as biographies of Pascal, da Vinci, Newton, and Kant.

In 1894 Filippov founded the journal *Nauchnoe obozrenie* (*Scientific Review*). Its contributors represented the broad range of Russian scholarship in the natural and social sciences and included the chemists Mendeleyev and N. N. Beketov, psychiatrist V. M. Bekhterev, clinician P. F. Lesgaft, and the leading Marxists Plekhanov and V. I. Lenin. Darwin, Helmholtz, Roentgen, and Marx were among the authors whose works were translated in its pages.

Filippov's first publication was entitled "Struggle and Cooperation in the Organic World" (1881). Darwin's great service, he explained, was his demonstration that relations among organisms had great significance for the variability of species. The investigation of these relations extended into the social sciences, and so Darwin's insight established him as "the founder of general sociology." Yet Darwin's contribution was compromised by the Malthusian assumptions that had obscured his understanding of the struggle for existence and blinded him to the importance of mutual aid. Struggle led to "the progress of some individuals," Filippov wrote, "but at the cost of regress for others." "Cooperation, on the other hand, constitutes the basis of harmonious progress," as was especially clear among humans.[67]

Filippov elaborated this argument in "Darwinism on Russian Soil" (1894). Here he criticized plant physiologist K. A. Timiriazev for his dogmatic defense of Darwin and for his evasiveness regarding the Malthusian origins of Darwin's theory.[68] For Filippov the very expression "struggle for existence" expressed the "one-sided Malthusian spirit" with which Darwinists approached nature. Blindly assuming that all relations among organisms were antagonistic, they failed to see "solidarity between organisms, which is no less important than struggle."[69]

Darwin's metaphor perpetuated Malthusian views by blurring the distinction between qualitatively different aspects of organic relations:

> In poetry it is entirely appropriate to speak of the struggle of a canoe with a wave, or of a sailboat with a storm; such expressions would also be permissible in biology if they did not cause confusion. The problem is that the term "struggle" often allows one to confuse direct adaptation to conditions with indirect adjustment, that is, with competition between organisms that have adapted in varying ways—and these are not at all the same thing.[70]

If a plant was unable to find water, it perished, irrespective of competition with others. Caterpillars did not compete with trees—they devoured them and depended on them for their existence. These were examples "of the mutual relations between entities for which Malthus's law, and consequently the selection theory, is inapplicable."[71]

Darwin's theory, then, was applicable only when organisms were in conflict. So, when examining nature in order to evaluate that theory, one should reserve the term "struggle for existence" for "such antagonistic relations between organisms according to which the life, development and reproduction of one is incompatible with [that] ... of another."[72] Such relations certainly existed, but they were not so prevalent as Malthus's "arithmetical trick" had indicated. By burying the "unscientific myth" of Malthusianism, Filippov sought also to inter "the basis of the struggle for existence in the sense of competition necessarily leading to natural selection."[73]

He did so in a manner familiar to Russian readers. First, he insisted that an organism's rate of reproduction responded to the influence of external conditions. Darwin's calculations regarding the length of time required for a pair of elephants to overpopulate the globe were abstract and irrelevant since "a gradual but significant lowering of fertility would occur long before any species" would do so.[74] Every botanist knew that the slightest environmental change elicited a dramatic change in a plant's rate of reproduction. It was even conceivable that under radically altered circumstances a dandelion might grow into a large tree by shifting resources from reproduction to growth.[75]

Second, Filippov argued that the central aspect of the struggle for existence was the organism's relationship to the physical environment. This, he charged, would have been obvious to Darwin had he spent less time in English gardens and more in the wild.[76] And had Darwin been less enamored of Malthus, he also would have appreciated a third point: that external conditions compelled organisms to rely on "collective struggle, giving birth to the solidarity of individuals united for a general goal."[77] This truth was manifest, to one degree or another, throughout the organic world:

> However great the struggle between people, one cannot forget that cooperation is also an essential aspect of their mutual relations. To a lesser degree the same is true of animals, especially higher ones, but also lower. Beginning with the phenomenon of so-called symbiosis ... and ending with the complex conditions of ... the lives of so-called social animals, we see an entire series of [cooperative] interactions among individuals.[78]

Cooperation so diminished intraspecific conflict that natural selection could play only a secondary role in evolution, which for Filippov resulted primarily from the direct action of the environment and the inheritance of acquired characteristics.

V. M. Bekhterev

Vladimir Mikhailovich Bekhterev (1857–1927) was a philosophical materialist, a populist, and a leading figure in Russian neurophysiology, psychology, and psychiatry from the late nineteenth century through the early Soviet period. From 1885 to 1893 he headed the Department of Psychiatry at Kazan University and organized Russia's first psychophysiological laboratory. In 1893 he moved to St. Petersburg, where he directed the Department of Psychiatry and the Department of Neuropathology in the Military-Medical Academy, founded and edited *Obozrenie psikhiatrii, nevrologii i eksperimental'noi psikhologii* (*Review of Psychiatry, Neurology and Experimental Psychology*) and, in 1909, organized a psychoneurological institute. His research encompassed the morphology, physiology, and pathology of the central nervous system; the localization of sensations; hypnosis; and psychotherapy. His doctrine of "reflexology" rested on his view that all psychological activities were complex reflex reactions.

In his article "Social Selection and Its Biological Significance" (1912), Bekhterev observed that Darwin's concept of the struggle for existence had never been "demonstrated scientifically." Darwin had arbitrarily placed this "terrible law at the basis of his discussions of mutual relations in the animal kingdom and, developing it as an indubitable truth, had arrived at his conclusion regarding the significance of so-called natural selection in the development of species."[79]

Bekhterev cited "our biologist Danilevskii," whose conservative political views were diametrically opposed to his own, to support his contention that intraspecific conflict was rare. The struggle of organisms against physical circumstances, on the other hand, was universal and unavoidable: "It should be obvious to anyone that what is universal is not the struggle for existence among individuals of the same species, or of different species, but rather struggle for the right of life generally, in other words, for the acquisition of the necessary conditions of existence from surrounding nature."[80]

When resources were inadequate to support all members of a species, conflict among them sometimes did ensue. More frequently, however, "more or less individual . . . adaptiveness often yields in importance to community work and social adaptiveness, since acquisition of the necessary conditions of existence, and defense from general dangers, often demands the joint forces of an entire series of individuals."[81] This tendency to join forces in the face of shared difficulties was manifest in the heightened sociability of organisms during the most difficult periods in their life, such as migration. It was evident even at the very lowest levels of the organic world and sometimes united members of different species, as "in the case of so-called symbiosis." Darwin had noticed that social selection was the basis of all community life, but had been blinded to its overriding significance by his principle of the struggle for existence.[82]

Nature's lesson for mankind was therefore clear. Humanity should encourage "those psycho-reflexes that support unity, mutual aid, and the division of labor," and would best flourish under the slogan "By common efforts for the general welfare."[83]

M. N. Bogdanov and His Tales for Children

Modest Nikolaevich Bogdanov (1841–1888) was a field zoologist with a lifelong interest in zoogeography, evolutionary theory, and the popularization of natural science. The son of a landowner in Simbirsk province, he attended the local gymnasium and entered Kazan University in 1858. There he received a candidate's degree for his work "Material for the Investigation of the Ornithological Fauna of Simbirsk and Kazan Provinces" (1866) and founded the Kazan Society of Naturalists in 1869. He then embarked on a two-year investigation of the region along the Volga from Kazan to Astrakhan and through the Caucasus.

After receiving his master's degree in zoology in 1871, he moved to St. Petersburg, where he served as curator of the Zoological Museum of the Academy of Sciences. He also embarked on numerous expeditions, including one with General K. P. Kaufmann's military expedition to central Asia (1873) and another, sponsored by the St. Petersburg Society of Naturalists, to the Aralo-Caspian region (1874).[84] During these and other expeditions, he collected specimens for a comprehensive catalogue of the birds of the Russian empire. He also studied the ornithological collections of Vienna, Stuttgart, Berlin, and Paris during trips to western Europe in the late 1870s. In 1881 he received his doctorate in zoology. His friend Kessler died that same year, and Bogdanov replaced him as professor of zoology at St. Petersburg University.

Bogdanov's publications concerned a wide range of zoogeographical and immediately practical subjects. In *Birds of the Caucasus* (1879), *Russian Shrikes and Their Kin* (1881), and *Essays on the Nature of the Khiva Oasis and the Kryzylkum Desert* (1882), he described and classified Russian wildlife and analyzed the relationship between organisms and their environment. He also founded the Russian Society of Poultry Raising, edited a journal on hunting and horse breeding, and spoke frequently to the Imperial Free Economic Society on such subjects as the elimination of gophers and the improvement of the cattle industry. His practical suggestions were often framed within general theoretical propositions. In an 1878 lecture on cattle breeding, for example, he sought to determine "the character of the variations of animal types under the influence of local, territorial and climatic conditions, in conformity with Darwin's well-known theory."[85]

Bogdanov read the *Origin* while a student in the gymnasium and always expressed great respect for its author. His initial enthusiasm for the selection theory waned, however, over the years. By the late 1870s he emphasized the direct action of the environment rather than natural selection, even when purporting to use Darwin's theory.[86] In his lectures on zoology in the academic year 1883–1884, he informed his students of two problems with Darwin's approach:

> Without disputing the fact that certain advantages may appear that give some individuals primacy over others in the life struggle we think, nevertheless, that one cannot unconditionally accept all of Darwin's propositions at the present time.

First, the struggle for existence and competition among individuals does not occur in the way that Darwin portrayed it; second, it is absolutely unthinkable that the organizational qualities conferring dominance in this struggle could appear accidentally and without cause, since everything in nature has a cause and nothing is accidental, exceptional.[87]

Bogdanov's changing attitude toward the struggle for existence can be traced through the fifty tales and essays about nature that he wrote for children between 1873 and 1888. Twenty-six were reprinted in the popular *Beggars of the Earth: Tales from the Life of the Animals Living around Man* (1884).[88] While on his deathbed Bogdanov extracted a pledge from his friend N. P. Vagner that the remaining works would be republished in another collection. Vagner kept his word; *From the Life of Russian Nature: Zoological Essays and Stories* appeared one year later.[89]

The titles of both collections reflected their self-consciously Russian character. "Beggars of the earth" is my own inadequate translation of the Russian "*mirskie zakhrebetniki*" (literally, "those on the spine of the world"). Bogdanov borrowed this title from a folk expression for those who were unable to support themselves and so "chew the bread of others and generally live on the world's back."[90] The title of the second collection, "From the Life of *Russian* Nature," reflected his desire to correct a deficiency in the stories available to Russian youths. These were translations of foreign works and simply did not ring true for Russian children. "Every people has its own culture, its own views of nature; every country has its own structure of life, its particular relations between man and animal. All this is unavoidably reflected in a story about nature and animals."[91] For example, the scarcity of wolves in Germany and England lent a certain "fairy-tale quality" to their portrayal by German and English authors—and this undermined their credibility for Russian children. Bogdanov offered tales drawn from Russian nature and unembroidered by "falsehood or fancy."[92]

His first story was entitled "Life on Earth (The Tale of an Elderly Mouse)" (1873). It reflected his enthusiastic acceptance of Darwin's theory and of a harsh, individualistic struggle for existence. A father introduces the tale by informing his son that organisms live entirely for themselves and that only the best adapted survive. "Everything ugly, everything weak, perishes at the first opportunity and is broken into pieces."[93]

A battle-hardened female mouse then picks up the narrative. Famine has precipitated a "fierce life-and-death battle" within her community. The stronger mice have preyed upon the "weak and small ones without the slightest twinge of conscience," littering the ground with the bodies of their victims.[94] Her siblings and father have perished, and her husband is hungrily eyeing their children—so she flees, seeking peace first in a field, then in a lake and a granary. The scene is always the same: "Wherever you turn—fighting and more fighting."[95] She finally returns home, where, at least, the enemies and their tactics are familiar.

The father then resumes the narrative, informing his son that this cruel struggle is all to the good. "When you become acquainted with the life of the animal world, not through your nanny's stories but in the field, the wood and the marsh, you will realize what a great role bitter battle plays in the progress of this world."[96] Competition has raised man above the animal world and should be encouraged since:

"The fiercer and more energetic the struggle, the more numerous its victims, the better—because only the tested warriors survive."[97]

These were rare sentiments for a Russian naturalist, and Bogdanov abandoned them shortly thereafter. The struggle of organisms against the elements and against other species remained central motifs in his later tales and essays, but intraspecific conflict vanished almost completely. Relations within a species were characterized by happy families with self-sacrificing mothers and by community cooperation in times of adversity. These themes became especially prominent after Kessler's speech "On the Law of Mutual Aid."

Typical of these later works was "The Autumnal Migration of Birds" (1884). As winter approaches, leaves and grasses begin to disappear, diminishing the food supply and leaving birds exposed to the "autumnal appetite" of their enemies.[98] Changing conditions compel a change in social relations:

> The birds, who had lived as individuals and quarreled with one another in the Spring, now meet as friends; old and young alike sense the need to live as a community. And so they collect into flocks. . . . It is as if they agree to earn their living together. . . . Like a regiment of soldiers on maneuvers they take disciplined flight along meadows and ravines. If you frighten the flock then one, another, a third bird, or all at once will rise and take flight in close order—where one goes, there go all.[99]

Cooperation was the key to success in interspecific conflict:

> When a bird is searching for food it does not see what is happening around it and so it is easy for an enemy to fall upon it. But if two, three or several birds gather together then some can search for food while others watch over them. Understandably, such a division of labor makes it much easier to protect oneself from danger. This is a very simple form of the mutual aid that we encounter among birds and many other animals.[100]

Experienced older birds led each flock, whose members were confident that "the leaders will save them from danger, from hunger, cold and hostile enemies."[101] Birds who attempted to migrate alone would almost certainly perish.

We do not know whether the dying Bogdanov remembered his first story, with its theme of incessant individualistic struggle, when he asked Vagner to republish his tales, nor do we know whether he would have included it in his final collection. Vagner may well have done so simply to keep his pledge. But the editors of the 1906 version of *From the Life of Russian Nature* felt no such obligation. Biologist V. A. Fausek and meteorologist P. I. Brounov reproduced the Vagner edition in all details save one: they excluded the mouse's tale. "This single story," they explained, "is distinguished by a one-sided and narrow tendentiousness." Without this discordant note Bogdanov's collection gained "simplicity and coherence of content."[102]

Kessler's Legacy: Mutual Aid and the Selection Theory

In itself the mutual aid tradition offered only an ecological conception, not an evolutionary theory. Proponents of the "weak" variant, such as Severtsov and Menzbir, identified no contradiction between the prevalence of cooperation in nature and a theory of evolution that relied on individualistic competition. But proponents of the "strong" variant, such as Kessler, Brandt, and Bogdanov, found themselves in a different situation. For them intraspecific competition was so vitiated by mutual aid that it could not play the central evolutionary role assigned to it by Darwin. Yet they could not simply replace intraspecific conflict with mutual aid to formulate a slightly reconstituted selection theory. Mutual aid provided an elegant explanation for anthills, flocks of migrating birds, and even ethics. Yet it did not explain the evolution of such physical characteristics as the stripes of a zebra or the shortened wings of a new species of insect.

So, on the one hand, mutual aid theorists usually praised Darwin for substantiating the fact of evolution and for enriching evolutionary theory with such valuable concepts as the struggle for existence and natural selection. On the other, their concept of cooperation in nature struck at the heart of the selection theory by negating the very process—intraspecific competition—that was Darwin's principal explanation for the divergence of characters and the emergence of new species. So when these mutual aid theorists joined their ecological conception to an evolutionary theory, they usually emphasized mechanisms other than natural selection. Most important were the direct action of the environment, isolation, and the inheritance of acquired characteristics.[103]

There was no debate among nineteenth-century Russian evolutionists concerning mutual aid and its relationship to Darwin's theory. One is struck by the fact that Russia's two leading exponents of "classical Darwinism," M. A. Menzbir and K. A. Timiriazev, criticized neither Kessler nor other mutual aid theorists in their many books and articles before the revolution of 1917. There were, I think, two related reasons for this. First, two years after Kessler's speech Tsar Alexander II was assassinated, ushering in a period of political reaction. This included a broad assault on science as a source of rational values, on evolutionism in general, and on Darwin in particular. In this context Menzbir and Timiriazev were preoccupied with the defense of a general position shared by mutual aid theorists. Second, Darwin's debt to Malthus had always made his theory suspect to many Russians, and Timiriazev was laboring mightily to overcome this association in the minds of his countrymen.[104] This effort would hardly have benefitted from an attack on naturalists who venerated Darwin and attempted to integrate his theory with a popular concept of cooperation in nature.

Throughout the nineteenth century, therefore, mutual aid remained an uncontroversial element in Russian evolutionary thought. It met a stiff challenge, however, when it was transported to a less receptive cultural milieu, Darwin's England, by the most famous heir to Kessler's legacy.

Thomas Robert Malthus (National Portrait Gallery, London)

Charles Darwin (permission of Darwin Museum and the Royal College of Surgeons of England)

Andrei Nikolaevich Beketov (Institute of the History of Natural Science and Technology, Leningrad branch, Academy of Sciences, U.S.S.R.)

A friendly caricature of Beketov
The Professor of botany: Do you know, children, what kind of tree this is?
Children: Betula alba, Professor!
The Professor: So take a look at it: be clever and study well!

Sergei Ivanovich Korzhinskii (Institute of the History of Natural Science and Technology, Leningrad branch, Academy of Sciences, U.S.S.R.)

Il'ia and Ol'ga Mechnikov in the 1870s (Pasteur Institute, Paris)

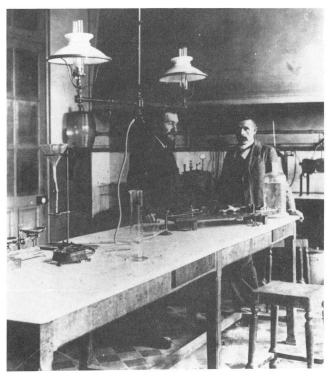

Mechnikov in his laboratory at the Pasteur Institute (Pasteur Institute, Paris)

A Parisian caricature of the "Apostle of Yogurt." The monkey and raven represent his preoccupation with evolution and death (Pasteur Institute, Paris)

Il'ia Il'ich Mechnikov, 1908 (Pasteur Institute, Paris)

Karl Fedorovich Kessler (Institute of the History of Natural Science and Technology, Leningrad branch, Academy of Sciences, U.S.S.R.)

Two bison graze in the Imperial Game Preserve, off-limits to Kessler and other naturalists (Library of Congress)

Petr Alekseevich Kropotkin as a military officer in Siberia, 1864

Kropotkin's sketch of a chase after horses in Siberia

The Amur River of Siberia. In the background, a military expedition travels on Bactrian camels (Library of Congress)

Kropotkin after his return to Russia (Institute of the History of Natural Science and Technology, Leningrad branch, Academy of Sciences, U.S.S.R.)

Nikolai Alekseevich Severtsov

Tsarist troops assemble at the gate of the palace of the Kokand khanate (Library of Congress)

The Kirghiz of Syr Darya with their domesticated goats (Library of Congress)

Kliment Arkadevich Timiriazev in Cambridge University robes in 1909

Timiriazev in his laboratory at Moscow University in 1898 (Institute of the History of Natural Science and Technology, Leningrad branch, Academy of Sciences, U.S.S.R.)

The Steppe in the Afternoon, oil on canvas by A. K. Stavrasov, 1852 (permission of The State Russian Museum, Leningrad)

CHAPTER 7

Kropotkin's Theory of Mutual Aid

Russian zoologists investigated enormous continental regions in the temperate zone, where the struggle of the species against natural obstacles (early frosts, violent snowstorms, floods, etc.) is more obvious; while Wallace and Darwin primarily studied the coastal zones of tropical lands, where overcrowding is more noticeable. In the continental regions that we visited there is a paucity of animal population; overcrowding is possible there, but only temporarily.[1] (P. A. Kropotkin, 1909)

Kropotkin considered himself a Darwinist and had every right to do so.[2] (M. A. Menzbir, 1922)

Prince Petr Alekseevich Kropotkin (1842–1921) is well-known to Western scholars as a leader of the international anarchist movement and as the author of *Mutual Aid: A Factor of Evolution*. One is tempted to see his ideas about cooperation in nature as the simple product of the strongly-held political convictions with which they were so compatible.

This temptation, however, reveals as much about the marginality of mutual aid theories in Europe and the United States as it does about the actual origin of Kropotkin's views. Kropotkin first questioned Darwin's approach to the struggle for existence while exploring Siberia as a youth and was an accomplished and celebrated naturalist years before his political views crystallized. Furthermore, as we have seen in Chapter 5, his ideas about cooperation in nature were quite common among Russian naturalists of varying political perspectives.

Kropotkin's theory of mutual aid certainly had an ideological dimension, but it cannot be dismissed as the idiosyncratic product of an anarchist dabbling in biology. What, then, are we to make of the fact that the most extensive expression of the Russian mutual aid tradition was written by an émigré anarchist in England? Mutual aid was not a controversial idea in Russia. Classical Darwinists there declined to attack it, nor did they associate Darwin's theory with the "Social Darwinist" doctrines that were so popular in the West. It was only when Kropotkin brought a Russian intellectual tradition into contact with a quite different English one that he felt compelled to elaborate what for many Russians was commonsensical.

The Anarchist Prince

A direct descendant of Russia's first rulers, Kropotkin was born in 1842. He matured in the heady atmosphere of the 1860s and shared with his older brother, Alexander, the common enthusiasm for natural science. The two discussed the scientific worldview emerging from A. I. Herzen's *Letters on the Study of Nature* and from the works of Vogt, Moleschott, and Büchner. By mid-1862 they had "devoured all the available books and articles on natural science, scientific materialism and political economy."[3] Petr heartily approved Alexander II's emancipation of the serfs in 1861 and, much moved by N. V. Shelgunov's account of poverty among western European workers, pronounced himself "an ardent defender of the proletariat."[4] These were not the sentiments of a convinced revolutionary, but rather of a sensitive young man influenced by the passions of his day and uncertain of his future plans.

He graduated first in his class from the Corps of Pages, and so became the tsar's *page de chambre*. A prosperous career in Alexander II's court beckoned, but Kropotkin soon found himself bored and disillusioned. He seized an opportunity to escape to the wilds of Siberia as naturalist on a series of commercial and military expeditions. Hurriedly consuming the works of such previous explorers as Richard Maack and Alexander von Humboldt, he departed St.Petersburg in June 1862.

In the next five years he traversed over fifty thousand miles, discovering en route an aptitude for geography and geology and receiving "a genuine education in life and human character."[5] The expeditions took him to Irkutsk and Chita, and then through eastern Siberia, along the Amur River to Manchuria. In 1865 he devoted himself exclusively to geography, studying the Ice Age formations of eastern Siberia in trips supported by the Imperial Russian Geographical Society. In 1866 he participated in the society's Olekmin-Vitim expedition, which sought an overland route for the transportation of cattle between Chita and the Lena gold mines. Throughout these years Kropotkin dispatched letters and scientific reports on his travels, and their publication gained him a certain celebrity.[6]

During the Olekmin-Vitim expedition he befriended I. S. Poliakov (1847–1887), the son of a Cossack. This former instructor at a military school in Irkutsk became the expedition's zoologist. Kropotkin took him under wing, worked closely with him during the expedition, and afterwards helped him prepare for entry into St. Petersburg University. They coauthored the expedition's scientific report, which was published in 1873, lived together briefly, and remained friends thereafter.[7]

While in Siberia Kropotkin was horrified by the harsh exploitation of workers in the Lena gold mines, struck by the futility of reformist efforts, and moved by dramatic instances of militant resistance.[8] He also met two famous revolutionaries in Siberian exile. One gave him a volume of Proudhon, which he read enthusiastically.

On his return to St. Petersburg he took an undemanding job with the Ministry of Internal Affairs, but concentrated on his promising scientific career. His widely acclaimed articles from Siberia gained him membership in the Imperial Russian Geographical Society. He was awarded a gold medal in 1868 for his work on the

Olekmin-Vitim expedition, and two years later became secretary of the society's section of mathematical and physical sciences. His subsequent publications, particularly on orography and glaciation, won him a considerable reputation in scientific circles.[9] During these years he befriended a number of experienced naturalists, including N. A. Severtsov, N. N. Mikhlukho-Maklai, N. M. Przheval'skii, and O. A. Fedchenko.

Kropotkin's political views apparently began to coalesce during 1870–1871. He frequented populist gatherings and met the editor of *Fatherland Notes*, N. K. Mikhailovskii, who was at the time writing about the consequences of Darwinism for the social sciences.[10] Inspired by the Paris Commune, Kropotkin became increasingly attracted to the International Workingman's Association, and in 1872 traveled to its headquarters in Switzerland. On his return to Russia he joined the Chaikovtsy circle and participated actively in its efforts to popularize revolutionary socialism among workers.

In March 1872 his two passions brought him both acclaim and trauma. After delivering a paper on the origins of the Ice Age, he was nominated by an appreciative audience to chair the Imperial Russian Geographical Society's section on physical geography. Hours later he was arrested and imprisoned in the Peter and Paul Fortress with other members of the Chaikovtsy, whose circle had been penetrated by tsarist agents. Influential friends in the society won Kropotkin permission to obtain the books and papers necessary to continue his scientific work in prison. He completed his *Investigations of the Ice Age* there in 1876.[11] That same year he escaped from a prison hospital and found refuge in England.

Between 1876 and 1883 Kropotkin emerged as a leader of the international anarchist movement. In late 1876 he met P. L. Lavrov, a fellow socialist émigré interested in developing a non-Malthusian evolutionism.[12] In the early 1880s he worked with Jean-Jacques Élisée Reclus, the French geographer, Communard, and anarchist, contributing his observations about Siberia to Reclus's massive *Nouvelle Géographie Universelle: La Terre et les Hommes* (1886). His continued interest in science was also evident in several contributions to the English journal *Nature*.[13] In December 1882 he was arrested in France and condemned to five years' imprisonment. This sentence was cut short by the authorities' decision, in 1886, that it was more expedient to expel the "anarchist prince" from the country.[14]

Making his home near London, he wrote prolifically about politics, history, and natural science. *In French and Russian Prisons* was published in 1887, followed by *The Conquest of Bread* (1892), *Fields, Factories, and Workshops* (1899), *Memoirs of a Revolutionist* (1899), *Modern Science and Anarchism* (1901), *Ideals and Realities in Russian Literature* (1905), and *The Great French Revolution* (1909).

Toward the end of his life, Kropotkin claimed, with but slight exaggeration, that he was better known in England as a scholar than as an anarchist.[15] He contributed regularly to *The Nineteenth Century* and in 1892 replaced the retiring T. H. Huxley as editor of its section "Recent Science." In the years 1892–1901 he wrote over fifty articles for the popular journal on a broad array of subjects, including artificial diamonds, light and electricity, the spectral analysis of stars, Antarctic explorations, stereochemistry, the experimental morphology of plants, muscles, heredity, and the evolution of the eye.[16] He also contributed over twenty articles on the geography of

Russia to the ninth, tenth, and eleventh editions of the *Encyclopaedia Britannica*.[17] His first article on mutual aid appeared in *The Nineteenth Century* in 1890; the others followed over the next four years. These were collected in *Mutual Aid: A Factor of Evolution* (1902), which appeared in Russian two years later. From 1905 to 1919 he produced a second series of articles in which he sought to establish a "true Darwinism" based on mutual aid, a reformulated concept of natural selection, the direct action of the environment, and the inheritance of acquired characteristics.[18]

A much-respected member of the scientific community, Kropotkin became a member of the British Association for the Advancement of Science in 1893, investigated remnants of the Ice Age on an expedition to Canada in 1897, and reported to the Royal Geographical Society in London on "The Dissication of Eur-Asia" in 1907.[19] Cambridge University offered him a chair in geology in 1896, but Kropotkin declined because the appointment was contingent on his abstention from political activity.

When the tsar was deposed in March 1917, Kropotkin returned to Russia but declined the offer of a post in the provisional government. Eight months later a second uprising brought power to the Bolsheviks, against whom he had polemicized vigorously. He continued to work on his final book, *Ethics* (1922), and on geology. In 1920 he declined an offer to teach geography at Moscow University, explaining that he did not want to move to the capital.

Kropotkin maintained an ambivalent attitude toward the Soviet government. On the one hand, he criticized Lenin for Bolshevik tactics during the civil war and thought centralized state socialism was unacceptably authoritarian. On the other, he preferred Soviet power to the likely alternative, restoration of the monarchy, and thought the Soviet state might, despite its negative aspects, prepare the necessary transition from capitalism to communism.

Kropotkin died in February 1921 and was honored by thousands at an enormous funeral.

The Siberian Experience

Like Charles Darwin's voyage on the *Beagle* (1831–1836), Kropotkin's travels in Siberia (1862–1867) exemplified what Susan F. Cannon has termed "Humboldtian science."[20] Darwin was twenty-two and Kropotkin nineteen when, as gentlemen-observers uncertain of their future plans, they toured unfamiliar regions, turning inquiring minds to the persons, places, flora, and fauna that they encountered.

The impact on each was profound and enduring. Darwin returned to England a committed naturalist already recognized for the observations and specimens he had gathered during his travels. Kropotkin returned to St. Petersburg having similarly earned public recognition and entrée into scientific circles by his dispatches from the frontier. Both Humboldtians soon consolidated their scientific credentials with well-received geological works based on observations from their journeys.

They had, however, seen quite different parts of the world. Darwin's voyage took him to many lands; some, like Patagonia, resembled the Russian plain,[21] but

England was a sea empire, and the voyage of HMS *Beagle* encompassed tropical regions as well. There Darwin explored tropical islands, where he gathered material that would provide important biogeographical evidence for his selection theory and wandered awestruck through tropical rainforests, where a rich flora and fauna thrived in a congenial and relatively stable climate. Russia, on the other hand, was colonizing a great continental expanse; military and commercial expeditions took Kropotkin to a seemingly endless Siberian wilderness, where organisms coped with a wide range of swiftly changing conditions, with drought and torrential downpours, with extreme heat and sudden blizzards. Kropotkin's trip differed from Darwin's in one other important respect: he had the benefit of the *Origin*, which he evaluated against the background of the Siberian wild.

It is difficult to establish the precise influence of such journeys on the subsequent ideas of participants. Careful scholarship, drawing on rich archival sources, has determined that Darwin's voyage shook his belief in the stability of species and provided a rich stock of experiences from which he drew while developing an alternative view. His stay in Bahia, Brazil, also inspired his description of nature as an "entangled bank" and perhaps contributed to Darwin's image of a plenitudinous nature in which organisms were packed tightly, like "wedges," into every inhabitable space.[22]

The influence of Kropotkin's Siberian experience on his scientific ideas has, of course, received less attention. Available evidence does not permit us to establish this relationship definitively. His scientific articles from Siberia concerned geology and geography, and in them he scrupulously deferred to Poliakov on zoological issues.[23] His letters from Siberia to his brother, however, contain tantalizing hints about his thinking, and these lend credence to Kropotkin's later recollection that Siberia had a decisive impact on his ecological and evolutionary views.

Petr and Alexander maintained a lively correspondence in the years 1862–1867. Alexander was the adored and demanding mentor, recommending books on philosophy, science, and political economy and impatiently criticizing his brother's reactions to them. At his suggestion Petr read the public lectures of the Russian evolutionist K. F. Rul'e; both brothers were almost certainly evolutionists before reading the *Origin*.[24]

Alexander was the first to mention Darwin, in a letter of March 1863. Darwin's theory illustrated the general truth that a scientific worldview need not violate one's poetic sense; indeed, it should enhance it. How "sweet" it must have been for Darwin to hit upon his conception of "the great Struggle for life in which all earthly organisms are involved"—an idea replete with the "sad glory of poetry."[25] Three months later Alexander sent Petr his own short essay on the origin of species, in which he expressed particular interest in the unresolved question of the causes of variations. In another letter he informed Petr of the imminent publication of Rachinskii's Russian edition of the *Origin* and predicted, in the spirit of 1860s scientism, that Darwin's work would prove "more dangerous [to the autocracy] than hundreds of A. Kropotkin's."[26]

Petr responded that he too had been thinking about evolution. For him the most vexing issue was the definition of "species." Until this was resolved the identification of transitional forms would be virtually impossible since these could always be designated separate species. Perhaps this issue could best be settled by an experiment in which animals that reproduced rapidly were subjected to slightly varied and gradually altered environments. Yet such an experiment would, of course, require "hundreds of years." In the meantime he was especially interested in the differences among the insects in various parts of the empire.[27] Petr described at length his love for the wild, harmonious beauty of Siberia, the product of "the eternal, indefatigable work" of natural forces.[28]

In one passage of this letter, he attempted to describe the Siberian scene in terms of Darwin's struggle for existence. He had difficulty doing so. We can almost hear him thinking aloud, attempting to describe the conflicts in Siberian nature in terms of Darwin's emphasis on competition among organisms. He informed Alexander that he was writing about

> that terrible struggle which each separate tree must sustain, and about the destruction, extermination to which the woods here are subject. Forest fires destroy them annually in great masses; water, storms and everything are against them—age-old forests are replaced by a young, small, thick grove, and many of these trees will also perish. But it seems to me that the struggle will be easier for the new trees—there will be fewer of them than there were of the large ones, but the forces of nature will not be so destructive of them. Or not....[29]

An emphasis on competition between life forms did not quite fit the Siberian scene. Petr may well have concluded that the physical forces at work were so overwhelming that the differences between the small and large trees could scarcely influence their chances for survival. He reminded Alexander again of the awesome power of the elements: "I have never seen such winds, such hurricanes (sandy ones) as here."[30]

In his articles for *Moskovskie vedomosti* (*Moscow Times*), Kropotkin portrayed Siberia as a sparsely populated land in which the sheer force of the elements dwarfed the paltry efforts of man and beast. For example, in an account of cattle raising on the steppe near Chindantsk station, he described the powerful snowstorms that suddenly swept the area. Horses were scattered by the wind; soggy and tottering, they either froze where they stood or sought shelter in a crevice or undulation, where they perished under piles of snow. Cattle weakened and lost their color. Despite their masters' most resourceful attempts to save them, they too died by the herd at such times.[31]

The most vivid passages in his later articles on mutual aid amplified this same impression of the raw power of the winds, rains, and snows of Siberia:

> The terrible snow-storms which sweep over the northern portion of Eurasia in the later part of the winter, and the glazed frost that often follows them; the frosts and the snow-storms which return every year in the second half of May, when the trees are already in full blossom and insect life swarms everywhere; the early frosts and, occasionally, the heavy snowfalls in July and August, which suddenly destroy myriads of insects, as well as the second broods of the birds in the prairies; the torrential rains, due to the monsoons, which fall in more temperate regions in August and September—resulting in inundations on a scale which is only known in America and in Eastern Asia, and swamping, on the plateaus, areas as wide as European States; and finally, the heavy snowfalls, early in October, which eventually render a territory as large as France and Germany, absolutely impracticable for ruminants,

and destroy them by the thousand—these were the conditions under which I saw animal life struggling in Northern Asia. They made me realize at an early date the overwhelming importance in Nature of what Darwin described as "the natural checks to overmultiplication," in comparison to the struggle between individuals of the same species for the means of subsistence.[32]

Kropotkin later recalled that, while in Siberia, he and Poliakov, "under the fresh impression of the *Origin of Species*,"

> vainly looked for the keen competition between animals of the same species which the reading of Darwin's work had prepared us to expect.... We saw plenty of adaptations for struggling, very often in common, against the adverse circumstances of climate, or against various enemies; ... we witnessed numbers of facts of mutual support, especially during the migrations of birds and ruminants; but even in the Amur and Usuri regions, where animal life swarms in abundance, facts of real competition and struggle between higher animals of the same species came very seldom under my notice, though I eagerly searched for them.[33]

Poliakov's publications provide indirect support for this recollection. In his report on the zoological results of the Olekmin-Vitim expedition, Kropotkin's friend and protégé emphasized organisms' constant "struggle with the cold," a struggle that became unwinnable when the snows fell. At such times herbivores were unable to find food and perished in great masses; hunters slaughtered the immobilized survivors by the hundred.[34] In "Family Habits among Aquatic Birds" (1874), Poliakov described intraspecific cooperation in the struggle against predators.[35] Three years later, in *Episodes from the Life of Nature and Man in Eastern Siberia* (1877), he described the awakening of "insect societies" during the April thaws, when they joined herds of pigs, dogs, wolves, ducks, geese, gulls, and other animals.[36]

These descriptive works lacked explicit theoretical content, but they lend credence to Kropotkin's recollection that the Siberian experience shaped his conception of the struggle for existence. They fully confirm his later comment that Poliakov had expressed their common impression in "many a good page upon the mutual dependency of carnivores, ruminants, and rodents."[37]

Kropotkin later claimed that Siberia influenced his reaction to Darwin's theory in four ways. First, the dramatically varying environmental conditions impressed upon him the centrality of the organism's struggle against abiotic forces. Second, he saw "paucity of life," rather than overpopulation, as a common condition in nature, predisposing him against the Malthusian elements of Darwin's theory.[38] Third, where animal life was abundant he saw colonies of rodents, herds of deer and other organisms, migrations of birds, and so on—in other words, mutual aid in the struggle against physical adversity and other species. Fourth, he observed that when physical conditions were harshest, and so intraspecific competition presumably most severe, the entire group's survival was imperiled. Survivors were few and "so much impoverished in vigour and health, that no progressive evolution of the species can be based upon such periods of keen competition."[39]

Kropotkin himself stressed that the conclusions he and Poliakov reached about the struggle for existence were common among Russians with field experience similar to theirs:

It is impossible to study like regions without being brought to the same ideas.... The same impression appears in the works of most Russian zoologists, and it probably explains why Kessler's ideas were so welcomed by the Russian Darwinists, whilst like ideas were not in vogue amidst the followers of Darwin in Western Europe.[40]

It seems likely, then, that Kropotkin's Siberian experience played a similar role in the development of his ideas about evolution and ecology as had Darwin's *Beagle* voyage in his. Just as Darwin came to doubt the fixity of species, so did Kropotkin, already an evolutionist, become skeptical of the importance of intraspecific competition. Just as the image of nature as an "entangled jungle" found its way into Darwin's *Origin*, so did that of nature as a Siberian expanse inform Kropotkin's essays on evolution.

Darwin stepped off the *Beagle* into a peculiarly British scientific and cultural milieu, one in which both he and Wallace would find Malthus's *Essay on Population* a reputable and inspiring source of ideas. Kropotkin, too, would eventually find himself in England—but as a visitor from a non-Malthusian culture.

Germination of Mutual Aid Ideas

If Kropotkin returned from Siberia skeptical about Darwin's emphasis on intraspecific competition, his doubts could only have been reinforced by life among the St. Petersburg intelligentsia. In 1868 the first Russian edition of Malthus's *Essay* appeared, with Bibikov's long introduction tying Darwin to Malthus and criticizing both for their one-sided analysis of the struggle for existence.[41] One year later Kropotkin's acquaintance N. K. Mikhailovskii used a review of the Bibikov volume to criticize the undemocratic, Malthusian aspects of Darwin's conception.[42]

Kropotkin's first published comment about Darwin, on the occasion of the naturalist's death in 1882, reflected the influence both of radical ideology and of the ideas about mutual aid that were current among nonradical Russian scientists. In a short article for the anarchist weekly *Le Révolté*, Kropotkin praised Darwin for demonstrating, first, that species were mutable and, second, that evolution resulted from the struggle for existence, natural selection, and the direct influence of the environment. Darwin had thus made a profound contribution to the spirit of "criticism and demolition" that promised to explode the religious and social fallacies of the age.[43]

The bourgeoisie, of course, attempted to fashion from the struggle for existence an argument against socialism. Yet it was clear, even if Darwin had not said so himself, that just the opposite was the case. "Others, applying his methods and developing his ideas" had proven that

> it is sociable species, where all individuals live in solidarity with one another, that prosper, develop and reproduce; while those which live by brigandage, like the falcon, for example, are decaying throughout the world. Solidarity and joint labor—this is what supports species in the struggle to maintain their existence against the hostile forces of nature. Far from excusing exploitation ... the investigations of Darwin and his successors comprise, on the contrary, an excellent argument to the

effect that animal societies are best organized in the communist-anarchist manner.[44]

Kropotkin thus treated his anarchist audience to a congenial, explicitly ideological perspective on the struggle for existence.

Yet this paragraph also reveals his debt to Russian naturalists of a decidedly moderate political perspective. That these Russians were the unnamed "others" who were developing Darwin's ideas in a proper direction is clear from Kropotkin's reference to falcons. Three years earlier Kessler had presented his "law of mutual aid" to an audience at the St. Petersburg Society of Naturalists that included Kropotkin's friend, zoologist N. A. Severtsov.[45] Severtsov had remarked that those falcon species that cooperated among themselves were more prosperous than even more physically gifted ones that did not.[46] The title of Kessler's address and a short summary of Severtsov's remarks appeared in the society's journal in 1880. Kropotkin followed Russian scientific journals closely and apparently saw this volume before writing his article for *Le Révolté*.

Kessler's speech itself appeared only later as a supplement to the same journal, and Kropotkin later recalled reading it in Clairvaux prison in December 1882 or early 1883. He did so with enthusiasm. Although he thought Kessler overestimated the part played by familial relations in the genesis of mutual aid, he welcomed his compatriot's central argument as an important correction to western European views. Encouraged by Kessler's speech and Russians' reaction to it, he began to collect material on mutual aid.

It was, however, T. H. Huxley's "atrocious article" "The Struggle for Existence in Human Society" (1888) that stimulated Kropotkin's major treatise on this subject.[47] Huxley did not espouse a triumphalist "Social Darwinism," but his rendering of the struggle for existence was Malthusian and pessimistic. His comparison of relations in the animal world with "a gladiator's show" could not have differed more dramatically from the view shared by Kropotkin and his colleagues in St. Petersburg.[48]

The same was true of Huxley's conclusions regarding the struggle for existence among humans. For Huxley humans had often attempted to escape the animal state, the Hobbesian war of all against all, and to put limits on the struggle for existence. Successes were but short-lived, doomed by "deep-seated organic impulses, which impel the natural man to follow his non-moral course."[49] Most important was the drive to reproduce. In times of abundance the rate of reproduction accelerated and inevitably reestablished the struggle for existence. Following Malthus, Huxley argued that even if a perfect society were established, and even if it were to achieve a perfect balance between population and food supply, "nonmoral Nature" would rend "the ethical fabric" with the birth of just ten new progeny.[50] "So long as natural man increases and multiplies without restraint," he reasoned, "so long will peace and industry not only permit, but they will necessitate, a struggle for existence as sharp as any that ever went on under the regime of war."[51]

Huxley replied to reformers and socialists, as had Malthus almost a century earlier, that greed, ambition, rivalry, poverty, and war were but the symptoms of "the deep-seated impulse given by unlimited multiplication."[52] These could not, therefore, be abolished by "fiddle-faddling with the distribution of wealth."[53] Rather, England could best attain relative prosperity by competing effectively in the international market, which required a low-wage, well-educated work force, and by keeping poverty at a level that did not threaten social stability.[54]

For Kropotkin, Huxley's article reflected a broad consensus among English intellectuals: "I found that the interpretation of 'struggle for life' in the sense of a war-cry of 'Woe to the Weak,' raised to the height of a commandment of nature revealed by science, was so deeply rooted in this country that it had become almost a matter of religion."[55] Only James Knowles, editor of *The Nineteenth Century*, and H. W. Bates, secretary of the Geographical Society, "supported me in my revolt against this misinterpretation of the facts of nature."[56] He responded hurriedly in articles for *The Nineteenth Century* on mutual aid among animals (1890), early peoples (1891), medieval city dwellers (1892), and contemporary societies (1894).

Mutual Aid: A Factor of Evolution

Like the Huxley article that had provoked it, *Mutual Aid* advanced a biosocial law based on an analysis of the struggle for existence. For Kropotkin natural selection generated a historical tendency toward cooperation in nature, including human society before the emergence of the modern state. In both realms mutual aid was the chief source of evolutionary progress.

His argument rested, not on the notion, which he associated with Rousseau and Büchner, that love was inherent to the natural world, but on an analysis of the dynamics of the struggle for existence. "No naturalist will doubt," he wrote, "that the idea of a struggle for life carried on through organic nature is the greatest generalization of the century. Life *is* struggle; and in that struggle the fittest survive." The central question was the relative weight of two different aspects of this struggle: "the direct one, for food and safety among separate individuals, and the struggle which Darwin described as 'metaphorical'—the struggle, very often collective, against adverse circumstances."[57]

This approach to Darwin's metaphor, which characterized *Mutual Aid* and all of Kropotkin's subsequent comments on evolution, contained an interesting "blind spot": he made no distinction between direct conflict and indirect competition within a species. Consider his recollection that while in Siberia he had not seen "a bitter struggle for the means of existence among animals belonging to the same species," although "I was eagerly looking for it." Indirect competition is not so readily visible. The ideological dimension to this is quite clear: for Kropotkin the central issue was whether intraspecific relations were essentially cooperative or conflictual, so the precise nature of any conflict was unimportant. The same might be said of Huxley's analogy between organic relations and a "gladiator's show"— the image to which Kropotkin was responding.

In *Mutual Aid*, Kropotkin introduced his argument through an analysis of Darwin's contradictory approach to the struggle for existence. On the one hand, Darwin's metaphor had originated in the "narrow Malthusian conception of competition between each and all"; on the other, Darwin had resisted the narrowness of

Malthus's formulation and had explicitly stated that he used this expression in a "large and metaphorical sense."[58] As a gifted naturalist Darwin was aware that natural selection had fostered cooperative relations among many species, giving rise to a social life and so to intellectual and moral faculties "which secure to the species the best conditions for survival." Thus, the concept of a struggle for existence "lost its narrowness in the mind of one who knew Nature."[59] Yet Darwin had been preoccupied with the consequences of individualistic competition and so had never pursued his insight about mutual aid. Furthermore, his own injunctions notwithstanding, the narrow Malthusian meaning of his metaphor had often reasserted itself in his work:

> Darwin never attempted to submit to a closer investigation the relative importance of the two aspects under which the struggle for existence appears in the animal world, and he never wrote the work he proposed to write upon the natural checks to over-multiplication, although that work would have been the crucial test for appreciating the real purport of individual struggle. Nay, on the very pages [of *Descent of Man*] just mentioned, amidst data disproving the narrow Malthusian conception of struggle, the old Malthusian leaven reappeared—namely, in Darwin's remarks as to the alleged inconveniences of maintaining the "weak in mind and body" in our civilized societies. As if thousands of weak-bodied and infirm poets, scientists, inventors, and reformers, together with other thousands of so-called "fools" and "weak-minded" enthusiasts, were not the most precious weapons used by humanity in its struggle for existence by intellectual and moral arms, which Darwin himself emphasized in those same chapters of *Descent of Man*.[60]

Kropotkin then reviewed Darwin's evidence for the prevalence of intraspecific conflict and found it astonishingly thin. The section of the *Origin* entitled "Struggle for Life Most Severe between Individuals and Varieties of the Same Species" contained

> none of that wealth of proofs and illustrations which we are accustomed to find in whatever Darwin wrote. The struggle between individuals of the same species is not illustrated under that heading by even one single instance: it is taken as granted; and the competition between closely-allied animal species is illustrated by but five examples, out of which one, at least . . . now proves to be doubtful.[61]

Darwin's principal argument for the prevalence of intraspecific competition, then, was the "arithmetical," or Malthusian, one. He "often speaks of regions being stocked with animal life to their full capacity and from that overstocking he infers the necessity of competition."[62]

For Kropotkin, of course, such Malthusian logic was unconvincing. In his political tracts he characterized Malthusianism as a pseudoscientific concoction expressing "the secret desires of the wealth-possessing classes."[63] In *Mutual Aid* he simply observed that it lacked explanatory value. For example, the population of southeastern Russia had remained stable for the past eighty years despite an average birthrate of sixty per thousand. A Malthusian might conclude that a "terrible competition between the inhabitants" had occurred, but in fact most newborns had died before "having grown to be competitors." The chief obstacle to population growth was environmental hazards (here, poor sanitary conditions) rather than intraspecific competition. The same was usually true among animals:

In the feathered world the destruction of the eggs goes on on such a tremendous scale that eggs are the chief food of several species in the early summer; not to say a word of the storms, the inundations which destroy nests by the million in America, and the sudden changes of weather which are fatal to the young mammals. Each storm, each inundation, each visit of a rat to a bird's nest, each sudden change of temperature, take away those competitors which appear so terrible in theory.[64]

This struggle against physical circumstances, and against other species, encouraged cooperation among like forms:

In the animal world we have seen that the vast majority of species live in societies and that they find in association the best arms for the struggle for life: understood, of course, in its wide Darwinian sense—not as a struggle for the sheer means of existence, but as a struggle against all natural conditions unfavourable to the species. The animal species, in which individual struggle has been reduced to its narrowest limits, and the practice of mutual aid has attained the greatest development, are invariably the most numerous, the most prosperous, and the most open to further progress.... The unsociable species, on the contrary, are doomed to decay.[65]

Kropotkin developed this argument by examining the combination of competitive and cooperative behavior among microorganisms, beetles, locusts, land crabs, termites, ants, bees, birds, beavers, rabbits, deer, antelope, gazelles, buffalo, goats, sheep, donkeys, horses, and finally—for most of his book—humans. Intraspecific relations contained elements of both competition and cooperation, the relative importance of which varied according to circumstances. Beetles, for instance, usually lived an isolated life, but when a beetle discovered the corpse of a mouse or bird the united efforts of many were required to transport and bury it. This was accomplished "without quarreling as to which of them will enjoy the privilege of laying its eggs in the buried corpse."[66] Among higher vertebrates mutual aid was "periodical, or is resorted to for the satisfaction of a given want—propagation of the species, migration, hunting or mutual defense."[67]

Although the relative importance of competition and cooperation fluctuated by season and circumstance, natural selection generated a historical tendency toward cooperation. "In the great struggle for life—for the greatest possible fullness and intensity of life with the least waste of energy—natural selection continually seeks out the ways precisely for avoiding competition as much as possible."[68] Mutual aid conferred an advantage in the struggle for existence by providing a better defense against predators and a more effective offense against prey, by conserving energy through joint efforts and by providing the sociable life necessary to the development of higher mental qualities.[69] For example, bees were capable of both sociability and predatory behavior toward their fellows. When there was no common work, "robbery, laziness and very often drunkenness become quite usual" among the bees populating sugar plantations. But when the harvest was rich, mutual aid prevailed. Clearly, bees always retained their "anti-social instincts," but their sociability was continually strengthened by natural selection since "in the long run the practice of solidarity proves much more advantageous to the species than the development of individuals endowed with predatory inclinations."[70]

Species that cooperated had a better chance of survival in the struggle for life than did less sociable ones, even if the latter were physically superior in every respect.[71] The decisive advantage of mutual aid was evident in the prevalence of cooperation, especially among higher organisms, and from the fact that "at the top of each class" of animals resided a highly sociable species.[72]

This tendency toward cooperation had been interrupted by human intervention. "When the Russians took possession of Siberia they found it so densely peopled with deer, antelopes, squirrels, and other sociable animals, that the very conquest of Siberia was nothing but a hunting expedition which lasted for two hundred years." There now remained only "the debris of the immense aggregations of old."[73] Human settlements had destroyed animals' food supplies and forced them to scatter in order to avoid intraspecific competition, consigning formerly gregarious creatures to an individualistic existence.[74] It was clear, for instance, that "during the last century the common weasel was more sociable than it is now."[75]

Domestication had the same result. In their natural habitat marmots were compelled to cooperate against abiotic forces and lived together in large groups characterized by "peace and harmony." Their dormant fighting instincts reappeared in captivity.[76] One could not, therefore, understand the struggle for existence in nature by studying domesticated animals. Darwin's attempt to illustrate the importance of intraspecific competition by examining the effect of drought on South American cattle was fatally flawed because wild "bisons emigrate in like circumstances to avoid competition."[77]

Mutual aid, then, dominated the natural world. Intraspecific conflict existed but had little evolutionary significance, except insofar as it led to migration and isolation.[78] Kropotkin endorsed Chernyshevskii's "remarkable essay upon Darwinism," with its thesis (formulated while Chernyshevskii lived in Siberia) that conditions harsh enough to stimulate intraspecific competition left survivors so weakened that any evolutionary progress was impossible.[79]

The same was true of humans. History testified to a constant struggle between tendencies toward competition and cooperation. As had become the fashion for naturalists, historians had "analyzed, described and glorified" the former while neglecting the latter.[80] Yet the "reckless competition for personal advantages" had proven less advantageous for the progress of human industry and ethics than had mutual aid.[81] Thus, cooperation had come to dominate daily existence and governance until the emergence of the modern state.

By becoming the repository for cooperative tendencies, the state had reversed the historical trend toward increasing cooperation among the citizenry. "The absorption of all social functions by the State necessarily favoured the development of an unbridled, narrow-minded individualism." Formerly, a bystander had felt obligated to interfere in a street fight for the good of all; in contemporary society he was encouraged to leave such matters to the police. "And while in a savage land, among the Hottentots, it would be scandalous to eat without having loudly called out thrice whether there is not somebody wanting to share the food, all that a respectable citizen has to do now is to pay the poor tax and to let the starving starve."[82] Thus, the tendency toward mutual aid, which constituted "the best guarantee of a still loftier evolution of our race," had been confined largely to the family, slum neighborhoods, the village, unions of workers, and the revolutionary movement.[83]

Despite the title of his work, Kropotkin devoted but few pages to mutual aid as a factor in the evolution of the physical characteristics that defined species. He dwelled on the advantages conferred by cooperation in the struggle for existence, on the consequent prevalence of mutual aid in nature, and on the necessary relationship between mutual aid, intelligence, and ethics—but he wrote very little about cooperation as a cause of speciation. A careful examination of his central argument—that "sociability appears as the chief factor of evolution both directly, by securing the well-being of the species while diminishing the waste of energy, and indirectly, by favoring the growth of intelligence"—reveals that, in itself, cooperation preserved rather than altered species.[84] Cooperative forms were able to survive and rear progeny with a minimal expenditure of energy, to maintain their numbers with a relatively low birthrate, to migrate more successfully, to develop greater intelligence, and generally to become more numerous, prosperous, and long-lived than species afflicted with intraspecific conflict.[85]

How, then, did mutual aid produce new species? Indirectly; by increasing the viability of a life form, it increased the possibility that another agent of evolution would act upon it. In one of the few cases in which Kropotkin considered the evolution of physical traits—a short discussion of squirrels—he attributed speciation to the direct action of the environment. He speculated that migration, a change in physical conditions, or acquisition of a new food supply might have a physiological effect on a number of squirrels. This would produce a new variety "without there having been anything that would deserve the name of extermination among the squirrels." "The intermediate links would die *in the course of time* without having been starved out by Malthusian competitors."[86]

As a description of animal ecology, then, mutual aid constituted a simple alternative to competition; as a factor of evolution it did not. This problem remained largely implicit for other Russian mutual aid theorists. Kropotkin, however, upheld this theory on the relatively hostile terrain of England. He felt compelled in subsequent essays to fully translate mutual aid from an ecological conception to an evolutionary theory; to demonstrate, as he expressed it in a letter to his fellow biologist and anarchist Marie Goldsmith, "that Mutual Aid does not contradict Darwinism, if natural selection is understood in the proper manner."[87]

Mutual Aid as a Theory of Evolution

Kropotkin anxiously monitored the reaction to *Mutual Aid* and wrote periodically to Marie Goldsmith about its reception.[88] He was especially proud of G. J. Romanes's estimation, in *Darwin and after Darwin*, that Kropotkin had presented "a large and interesting body of facts showing the great prevalence of the principle of cooperation in organic nature."[89] In a revised edition of *Animal Behavior* C. Lloyd Morgan had also conceded the prevalence of mutual aid in nature but had questioned its significance as a factor in evolution. Kropotkin informed Goldsmith that Morgan had insisted, "as befits a Darwinist/versus Lamarckian," that individual development was central to evolution.[90]

Kropotkin usually responded testily to criticism but agreed that Morgan had a point: mutual aid did not in itself provide a satisfactory alternative to evolution through intraspecific conflict. In his columns on "Recent Science" for *The Nineteenth Century* in 1893 and 1901, and especially in the articles "The Theory of Evolution and Mutual Aid" (1910), "The Direct Action of Environment on Plants" (1910), "The Response of the Animals to Their Environment" (1910), "Inherited Variation in Animals" (1915), and "The Direct Action of Environment and Evolution" (1918), he attempted to develop a non-Malthusian Darwinism that incorporated mutual aid and other factors that had come to be characterized as Lamarckian.[91]

In "The Theory of Evolution and Mutual Aid," Kropotkin addressed the apparent tension between his and Darwin's theories. Western Darwinists, he observed, were leery of the theory of mutual aid because it seemed incompatible with "the hard Malthusian struggle for life which they consider as the very foundation of the Darwinian theory of evolution." The authors of the selection theory had, after all, emphasized "the *individual* Malthusian struggle for *individual* advantages in their theory of Natural Selection."[92]

> It is useless to deny that a certain contradiction exists. If a strenuous Malthusian struggle for food and for the possibility of leaving progeny is carried on within each animal group to the extent admitted by most Darwinists (which *must* be admitted if the natural selection of individual variations plays the part that is attributed to it), then it excludes the possibility of association being a prevalent feature among animals. And, *vice versa*, if association prevails in the animal world to the extent we see it depicted in the works of our best field zoologists—the very founders of descriptive zoology: Pallas, Azara, Rengger, Audobon, Naumann, Prince Wied, Brehm, &c. . . . then struggle for life cannot possibly have the aspect of an acute inner war within each tribe and group. It cannot be a struggle for *individual* advantages. It must be an *associated* struggle of the group against its common enemies and the hostile agencies of environment. Natural selection in this case also takes a quite different aspect.[93]

He resolved this contradiction by redefining the role and scope of natural selection. Evolution occurred as follows: Organisms engage in a constant struggle for existence, primarily against abiotic conditions but also against other species. In the course of this struggle they tend increasingly to practice mutual aid. Individualistic species tend to decay, while those that cooperate increase in number and available energy. Thus, cooperative species are more subject to the direct action of the environment, which produces directed variations and is "the main factor of all evolution." These variations are passed to the next generation through the inheritance of acquired characteristics.[94] Natural selection, then, "ceases to be a selection of haphazard variations . . . but becomes a physiological selection of those individuals, societies and groups which are best capable of meeting the new requirements by new adaptations of their tissues, organs and habits." It operates largely "as a selection of groups of individuals, modified all at once, more or less, in a given direction." Understood in this manner, natural selection always favors the restriction of internal conflict and operates most forcefully when conditions are bountiful, since this lends "a certain plasticity to organisms."[95]

Kropotkin's reliance on the direct action of the environment led him to label "the physiology of variability and, consequently, of evolution" as the "task of the day." As he confided to Goldsmith in 1911, "Anatomy, histology and the changes occurring *in the cells*—always in the sense of adaptation: this is most interesting and in it lies the key to evolution."[96]

His correspondence with Goldsmith also reveals his a priori attachment to another essential element in his evolutionary theory, the inheritance of acquired characteristics. While readers of *The Nineteenth Century* perused his confident summaries of the "excellent works" on this subject by Paul Kammerer, Ludwig Plate, Richard Semon, and others, Kropotkin was oscillating between certainty and desperation. In December 1909 he condescendingly "congratulated" Goldsmith for being "somewhat more skeptical than necessary regarding the possibility of the inheritance of acquired characteristics." He was himself "profoundly convinced" that such inheritance was frequently affected through changes in the blood and other fluids that nourished the "spermplasm."[97] On March 8, 1910, however, he admitted that his examination of about two hundred works on the subject had uncovered "very few, by either botanists or zoologists, which would speak persuasively to even an inattentive reader." Did Goldsmith know of any "*good* works, which give the results of obvious variations—for example, in an underground laboratory—and where it turned out that variations *were* inherited?" He hastened to add that she need not search too vigorously, since "of course" he had already collected a few "convincing cases."[98] In August 1910 he lamented the limitations and anecdotal quality of the available evidence and complained that precise experimental studies were urgently needed.[99] In 1913 and 1915 he praised Semon's investigations; these, he claimed, had enabled him to make a "good argument in favor of the inheritance of the results of use and disuse."[100]

Back to Darwin!

Kropotkin greatly admired Darwin. Each of his essays about evolution began with a discussion of Darwin's views, and in each he presented his theory as the logical conclusion of Darwin's own theoretical development. For Kropotkin, Malthus's influence had, of course, led Darwin to certain erroneous views. Yet, as was clear in the sixth edition of the *Origin* and in Darwin's recently published correspondence, the great naturalist had been approaching a theoretical position much like that which Kropotkin now espoused.

Whig history it surely was, but Kropotkin's account of Darwin's changing views will not strike modern historians of biology as entirely fraudulent: When formulating his theory of evolution, Darwin sought primarily to prove that species were mutable and to rid biology of teleological thinking. The concept of natural selection allowed him to do so, and Darwin, feeling "a sort of paternal predilection" for this mechanism, "unduly minimized the direct action of surroundings upon living beings."[101] Concerned that his theory would suffer through association with Lamarck's, Darwin took pains, in the first edition of the *Origin*, to emphasize the differences between his ideas and those of the French transmutationist.[102] After pondering criticisms of the *Origin*, Darwin recognized that he had underestimated the importance of the direct action of the environment. Conceding his error in a letter to Joseph Hooker in June 1860, he rectified it in *The Variation of Animals and Plants under Domestication* and especially in the sixth edition of the *Origin*.[103] No reader of these works could doubt the mature Darwin's "Lamarckism."[104]

Darwin's approach to the struggle for existence had undergone a similar development. Having always recognized the complexity of this struggle, he had postponed the task of discriminating between its different aspects, "so widely different as to their consequences for the genesis of new species and for natural selection altogether."[105] Darwin's published letters revealed, however, that he had not been blinded by the Malthusian metaphor. While preparing his never-completed volume on natural selection, he had, for example, attempted to gather information about birds

> in order to see whether the majority of the arithmetically computed competitors which disappear every year are not destroyed already in the eggs, or as fledglings, so as to deprive the struggle for life of its competitive character and render it entirely metaphoric. Unfortunately, he never terminated this part of his researches.[106]

Darwin had, however, lived to amend his theory in another important manner: by attributing a greater evolutionary role to isolation, he had eliminated the need for evolutionists to postulate an incessant struggle among organisms.[107]

Evidence substantiating the evolutionary significance of isolation, the direct action of the environment, and the inheritance of acquired characteristics had continued to accumulate in the years after Darwin's death. These factors offered a substitute for "an acute struggle between the individuals of the species to preserve the effects of variation. The acting cause will itself accumulate them, and increase them in the subsequent generations."[108] This did not signal the demise of natural selection, and certainly not of the struggle for existence, but only of the Malthusian elements in evolutionary theory.

Several "considerations lying outside the true domain of biology," however, had prevented evolutionary theory from progressing in the ideal, logical fashion indicated by Darwin's own intellectual development and by the evidence accumulated since his death.[109] First, Darwin's successors were blinded by a Malthusian orthodoxy. Second, they were committed to a false polarization into "Darwinist" and "Lamarckian" camps, each of which distorted the positions of their namesake in a manner consonant with the fashionable idealism of the time. Third, biologists had lost touch with the naturalist tradition and so with nature itself.

Kropotkin railed constantly against these developments. The "vulgarizers of the teachings of Darwin," he wrote in 1918, "have succeeded in persuading men that the last word of science was a pitiless individual struggle for life." Yet mounting evidence about alternative evolutionary factors was undermining "the Malthusian idea about the necessity of a competition to the knife between all the individuals of a species" and was preparing "a quite different comprehension of the struggle for life, and of nature altogether."[110]

The relative lack of field experience among contemporary biologists also sustained the "Malthusian idea." In *Mutual Aid* Kropotkin had argued that the centrality of the organism's struggle with the environment, and the importance of mutual aid to this struggle, could be fully appreciated only by observing animals, "not in laboratories and museums only, but in the forest and the prairie, in the steppe and the mountains."[111] Writing to Goldsmith in 1909 about one critic of *Mutual Aid*, he attributed rejection of his views to "colossal ignorance" of the actual life of animals:

> In my conclusions I can err, I could have exaggerated the significance of sociability in the animal world (in the excellent company of others,—with the Darwin of the *Descent of Man*!). But the facts are not mine, [but those of] the fathers of descriptive zoology who, in the first 1/2 of the 19th century, studied the life of animals when, still practically untouched by man, it swarmed over the prairies and woods of both Americas.[112]

Surely, "if Audobon, d'Orbigny, etc. were alive they would have immediately introduced this correction into Darwinism."[113]

The progress of evolutionary theory had also been inhibited by the rise of philosophical idealism, by the lamentable tendency of contemporary intellectuals to "reject the healthy materialism of the natural sciences." This was a symptom of profound social crisis, "a reflection of the inclination toward pessimism which everywhere characterizes the day, especially in Germany—a time of well-known decadence, which enjoys everything mystical and fears all that is exact, like the [18]30s, a time of romanticism and naturphilosophe."[114] The vogue of idealist metaphysics, of the "childish jabbering" of Mach, Ostwald, and "this tin god Bergson," extended also to evolutionary theory.[115] He reminded Goldsmith:

> You saw to what extent this is a Europe-wide disease when you were writing about Darwinism. Look at what the neovitalists ... make of Lamarckism. The priests began this campaign in the 1860s, and conduct it intelligently, subtly, maliciously. They began with "uninvestigated phenomena," that is, with floating tables and hysterical women, and have concluded with this [idealist current] that has conquered the universities.[116]

In the same vein Kropotkin applauded the useful "fundamental work devoted to the general position of Darwinism and Lamarckism" by Hugo de Vries, Edward Drinker Cope, Vernon Kellogg, Ludwig Plate, Yves Delage, and Marie Goldsmith, but complained that biologists had been too often distracted by metaphysical speculations about heredity:

> So many side-issues have been introduced into the subject of Evolution, and so much that is purely dialectic has been dragged into the discussion, that a substantial portion of these works is given to heredity and the discussion of the rapidly altered and readjusted hypotheses of Weismann and his "Neo-Darwinist" followers, as also to the sudden "mutations" observed by de Vries, in which some naturalists saw a rival to, and some others a support of, Darwinism; to the rules of hybridism discovered by Mendel; and to the half-mystical reasonings of Pauly. In some modern works ... one will find, by the side of serious scientific analysis, a large tribute paid to the metaphysical and half-mystical views of some German biologists who try to revive the Hegelian *Naturseele*, in order to explain evolution; and most improperly describe themselves as "Neo-Lamarckians."[117]

Kropotkin identified this idealist current in the neo-Darwinian camp with August Weismann, whom he characterized in 1901 as "the Karl Marx of Biology, just as superficial, making grandiose generalizations upon a handful of facts—

metaphysics on a foundation that will not stand up."[118] Weismann was the object of numerous barbed comments in Kropotkin's essays of 1901-1918.[119] Weismann's doctrine was of theological origin, Kropotkin argued; his "Hegelian conception of an 'immortal germplasm,'" this "matter endowed with an immortal soul," was but an attempt to "reconcile teleology with mechanism." Weismann's deductive method contrasted sharply with Darwin's commitment to induction; indeed, it constituted a return to a pre-Darwinian teleology.[120]

Idealism had also polluted the Lamarckian well. Vulgarizers and unfriendly commentators had inaccurately branded Lamarck as a believer in evolution through volitional adaptation.[121] Not only had this led Darwin, in the first edition of the *Origin*, to portray Lamarck as "little short of an idiot," but it had also provided an opening to latter-day idealists.[122] So contemporary "Lamarckians" distorted Lamarck in the same idealist spirit as "Darwinists" distorted Darwin. Kropotkin informed Goldsmith in 1909 about his current work:

> I have here clashed again with Lamarckism ... all the more since now ... an entire orientation is beginning to emerge: to employ the Lamarckian direct action of the environment, or more accurately, as it is expressed, the organism's adaptive capacity, to strike at Darwinism and to return to some sort of *force directrice de l'évolution* and, in the final analysis, to the Hegelian Universal Spirit and such nonsense.[123]

There was, Kropotkin argued, a simple solution to the crisis of evolutionary theory: "Back to Darwin!"

> But to the Darwin of 10 years after the appearance of his Origin of Species when, reinforced by the new science that he himself had elicited and by the immense research which he had conducted and systematized on variation, he was able to provide a theory of evolution in which natural selection was reduced to its true role of a lady which chooses the clothing which has been offered and prepared by her *camériste*, by variation; but she *must* accept when they are the product of new conditions of existence in the environment—a theory that ... [recognizes] the direct action of the environment as the *only* means of explaining *determined* and *adaptive* variation. A return, then, to a theory of evolution which is a synthesis of the "French theories" ... and [Darwin's] important observations on the struggle for life, which evidently is an important factor of evolution, but which should not be exaggerated to the degree it has been, contrary to Darwin's wish, by the phony Darwinists. This is a theory of evolution which, following Bacon, recognized the importance of Mutual Aid—that is, of the social instinct—for the preservation of species, and which, with Bacon, saw in it the primordial element of Ethics. ... This is above all a return to the Darwinism which saw in Evolution a spontaneous result of the forces of Nature, and not, as Weismann and his disciples wished, an evolution *predetermined* (by the Mechanisms of the Universe) by means of a substance possessed of an "immortal" soul—this Hegelian creation of Weismann, his germ plasm. This [germ plasm theory] ... has plunged biology into a situation where, for 30 years, hundreds of pages of dialectical biology have been written in place of the experimental research of previous times. A theory, finally, of Evolution [which follows Herbert Spencer, Auguste Comte, Claude Bernard, and G. H. Lewes by including] ... a *physiological evolution of organs* [*caused*] *by the new functions* which they perform as the organism is placed in new conditions of existence.[124]

Kropotkin's Darwinism, then, was an evolutionism in which isolation, the direct action of the environment, and the inheritance of acquired characteristics filled the explanatory void created by the bankruptcy of Malthusian postulates, postulates that the mature Darwin had himself come to question.

Geography, Ideology, and Evolutionary Theory

Kropotkin's theory of mutual aid was integral to his political outlook, and he considered an individualistic rendering of the struggle for existence to be equally integral to the bourgeois worldview. It is especially interesting, therefore, that in both published articles and private correspondence he attributed his differences with Darwin and Wallace, not primarily to their divergent ideologies, but to their differing experiences in the field. Kropotkin saw himself as part of a larger tradition common to Russian naturalists of various ideological orientations. Pondering criticisms of his evolutionary theory in a letter to Goldsmith in 1909, he observed that "Kessler, Severtsov, Menzbir, Brandt—4 great Russian zoologists, and a 5th lesser one, Poliakov, and finally myself, a simple traveller, stand against the Darwinist exaggeration of struggle within a species. *We see a great deal of mutual aid*, where Darwin and Wallace see *only struggle*."[125] Not one of the other Russians mentioned were politically radical. Kropotkin attributed their shared differences with Darwin and Wallace to the physicogeographical circumstances of their homeland:

> Russian zoologists investigated enormous continental regions in the temperate zone, where the struggle of the species against natural obstacles (early frosts, violent snow-storms, floods, etc.) is more obvious; while Wallace and Darwin primarily studied the coastal zones of tropical lands, where overcrowding is more noticeable. In the continental regions that we visited there is a paucity of animal population; overcrowding is possible there, but only temporarily.[126]

Here Kropotkin exaggerated the proportion of Darwin's field experience that transpired in crowded tropical settings, but this only underlines two points of interest to us. First, given Kropotkin's great respect for Darwin and Wallace, he preferred to attribute their errors to their specific experiences as naturalists rather than to their ideological prejudices. Second, for Kropotkin, Darwin's portrayal of nature, with its wedgelike packing of organism against organism, was clearly a description of the tropics and contradicted his own memories of Siberian nature.

These memories dominated his last recorded comments about nature and society. In 1920, when pondering the Bolshevik revolution and the political options available to Russian anarchists, the dying Kropotkin drew on images borrowed from his youthful travels through Siberia. The revolution resembled a typhoon buffeting the shores of eastern Asia, destroying ships and homes by the hundred. It was a "great natural phenomenon, a great catastrophe which shall either renew, or destroy; or perhaps both destroy and renew." As for the political options, these were few and almost irrelevant amid the elemental forces at work. Whether struggling amid a Siberian blizzard or a social revolution, a single individual or even small groups of individuals "not forming a fairly large mass are undoubtedly powerless—their powers are certainly nil."[127]

CHAPTER 8

Severtsov, Timiriazev, and the Classical Tradition

> Darwin suddenly posed an unexpected question. "Tell me, why do these German scholars quarrel so among themselves?" "You would know better than I," was my reply. "How so? I've never been to Germany." "Yes, but this is only further confirmation of your theory: they must have bred too much. It is yet another example of the struggle for existence." He hesitated for a moment and then broke out into the most good-hearted laughter.[1] (K. A. Timiriazev, on his visit to Down in 1877)

Several important Russian naturalists found Darwin's approach to the struggle for existence unobjectionable and even useful. Two of these exceptions to the Russian rule were N. A. Severtsov and K. A. Timiriazev. These two men differed markedly in temperament, experience, and career pattern; in quite different ways they serve to illuminate the pattern that they violated.

Severtsov was a naturalist-pioneer who spent most of his adult years with the tsarist military forces that were colonizing central Asia. He never held a university chair and gained a reputation among his urban colleagues as a romantic and exotic figure. More comfortable in the mountain ranges of Western Turkestan than the parlors of St. Petersburg and Moscow, he could scarcely have been more distant from the circles, journals, and ideological debates that shaped the outlook of urban intellectuals. Not once in his published work did he even mention the Darwin-Malthus connection. An evolutionist before 1859, Severtsov was predisposed by his own research to accept Darwin's emphasis on intraspecific conflict and found Darwin's "struggle for existence" a useful metaphorical shorthand for natural relations. His zoogeographical approach to issues in evolutionary theory became one source of Russian classical Darwinian thought.

Timiriazev, on the other hand, was a laboratory scientist, a professor at Moscow University, and a central figure in the urban intellectual scene. A product of the tumultuous 1860s, he considered scientific research and propaganda of its findings to be the pressing tasks of the age. This leading academic opponent of tsarism was also Darwin's unwavering "Russian Bulldog." Opposed to Malthusianism as a political doctrine, he nevertheless defended Darwin's use of Malthus's insight into population and food supply, which was consonant with his own work on the conservation of energy and plant physiology. A committed and prolific defender of Darwin's theory for over fifty years, Timiriazev knew his audience well. His changing rhetorical tactics, therefore, provide a sensitive indicator of Russian resistance to Darwin's metaphor.

N. A. Severtsov

> He reviewed and examined Darwin's theory for many years, and did so not through books, not in a study, but in the vast mountainous spaces of central Asia.[2]
> (M. A. Menzbir, 1886)

Nikolai Alekseevich Severtsov (1827–1885) was born in Voronezh province on the estate of his father, a discharged army officer who had lost a limb in the battle against Napoleon's army at Borodino. He was extensively tutored. By the time of his departure for Moscow University in 1843, he had gained a facility with German, French, English, and Latin, studied Buffon's *Histoire naturelle*, and, under the tutelage of the accomplished hunter F. I. Laushen, learned to observe carefully and shoot accurately.[3]

At the university he studied with evolutionist K. F. Rul'e.[4] Rul'e had recently revised the curriculum in zoology, deemphasizing descriptive studies and concentrating on the general laws of "animal organization and life (zoobiology)."[5] He encouraged Severtsov to study the fauna of Voronezh, which he did in annual trips beginning in 1844. By the time this project ended, he had spent over forty months living in the wild and observing the migration, moulting, and morals of Voronezh's wildlife.

These he described in his master's thesis, *Periodic Phenomena in the Life of the Beasts, Birds and Reptiles of Voronezh Province* (1855), which created a sensation among naturalists and earned Severtsov the Academy of Science's Demidov Prize. One of the academy's judges, A. F. Middendorf, praised *Periodic Phenomena* as "a completely new phenomenon in Russian literature" by its demonstration of "the tight connection between the distinguishing characteristics of a given fauna and the particularities of the soil upon which it lives and the climate in which it breathes and develops."[6] Beketov also praised Severtsov's treatise effusively and commented that its wealth of detail could be put to good use by hunters.[7]

Such acclaim did not, however, win Severtsov a university appointment. His application for a position as dozent in zoology was deferred by a Moscow University committee for fear that he would make an ineffective teacher.[8] By summer 1856 he had lost hope; he left Moscow for the museums and zoological gardens of western Europe. There he researched and wrote a long essay on the evolution of cats. A thirty-five-page summary of his conclusions appeared in the Parisian *Revue et magasin de zoologie pure et appliqué* in 1857 and 1858.

The rebuff by Moscow University proved a turning point in Severtsov's career. He gravitated toward an alternative source of support, the Russian military, and an alternative way of life, that of the field naturalist. The expansion of the tsarist empire into central Asia was opening up new markets, new sources of raw materials, and new areas of exploration for European naturalists. In 1847 the Russian military had begun a campaign against the Kokand khanate, which had become an important exporter of raw cotton and silk, as well as a base for raids into areas under tsarist control.[9] In 1865 the territories occupied by the tsarist military were united into the province of Turkestan under Governor-General Cherniaev, who became Severtsov's first military patron. Cherniaev's successor, K. P. Kaufmann, expanded the empire to the borders of Afghanistan and also monitored Severtsov's scientific work.

For most of the years 1857–1879, Severtsov traveled throughout Western Turkestan under Cherniaev's and Kaufmann's commands. Aside from pursuing the zoogeographical research that was his abiding passion, he provided geographical and geological information useful for the incorporation of the captured territories into the empire and for the extraction and transportation of their mineral resources. He also assumed duties of "a purely military character," serving occasionally as Cherniaev's chief of staff and helping tsarist forces negotiate their way through difficult terrain. Once, in later years, he was chosen as a truce envoy to Iakub khan after two previous emissaries had been impaled on stakes.[10] His enthusiasm for Russia's colonizing mission was evident in such articles as "On Places in Tian Shan Suitable for a Russian Population" and in several elaborate proposals that he submitted to Governor-General Kaufmann. These included plans for privately owned steamship lines and railless locomotives to facilitate trade, as well as a scheme for altering the course of several regional waterways. Kaufmann rejected each as impractical.[11]

Severtsov's adventures in central Asia soon brought him great fame. The Academy of Sciences sponsored his first trip to central Asia in 1857, where he and botanist I. G. Borshchev were to investigate "the influence of an extreme continental climate on plant and animal life."[12] The following year he was captured after a skirmish with the Kokands. According to Severtsov's celebrated account of the episode, his assistant was shot and killed, and he himself was subdued only after being stabbed, hit in the head with a sabre, and stunned by the explosion of his own rifle. He remained a prisoner for one month, during which time he found great comfort in prayer and "learned from experience the salutary significance of religion." Finally, three hundred Kazakhs, supported by two cannon, won his release.[13] On his return to St. Petersburg in late 1859, "the story of his captivity was in everybody's ears, and he had become the hero of the day."[14]

Capitalizing on his fame, Severtsov announced in February 1860 that he would deliver four public lectures to benefit needy students. His subject was the variability of species, with an emphasis on the role of theory in scientific cognition. "Theories glitter and vanish like soap bubbles," he wrote in his announcement, "but one cannot therefore reject them, as does the strictly descriptive school of zoology, for without ideas facts are also inaccessible to us."[15]

The occasion was irresistible and the hall packed. The public flocked to see this survivor of Kokand imprisonment, this celebrated, if forgetful, zoologist, subject of "innumerable stories about his legendary absentmindedness and forgetfulness."[16] Severtsov proved an inept lecturer. He arrived at eight o'clock, one hour late and poorly prepared. After reading his remarks for several hours, he announced, with no break in his monotone, that "now we will take a look at the clock"—and did so unremittingly until completing his lecture at eleven-thirty. By this time the hall was almost empty. The remaining three lectures were canceled.[17]

Shortly thereafter he departed for a two-year trip along the Volga and Ural rivers, where he studied the fishing industry and served on the Committee for the Building of a Ural Cossack Army. After a brief trip abroad he joined Cherniaev's detachment to Tian Shan in Western Turkestan.

He returned briefly to St. Petersburg in 1864, where he read Rachinskii's edition of the *Origin*, declined Kiev University's offer of a professorship, and prepared for another trip to Turkestan with Cherniaev's campaign against the Kokands.[18] In the years 1865–1867 he collected thousands of specimens of Turkestan fauna, mostly birds. On one expedition he met and married S. A. Poltoratskaia, a collector of insects and plants. Their son, the future zoologist A. N. Severtsov, was born in 1866. The following year Severtsov was awarded a gold medal from the Imperial Russian Geographical Society for his reports on the mountains of central Asia, and Moscow University awarded him a doctorate in zoology.

Support from the military and the Imperial Russian Geographical Society provided unprecedented access to the fauna of central Asia, but it also inhibited Severtsov's pursuit of the theoretical issues that most interested him. Governor-General Kaufmann proved a generally sympathetic patron, but he was less interested in refinements of evolutionary theory than in precise information about the geology and geography of the captured territories. He encouraged his charge to think less of the former and produce more of the latter, underlining this request on one occasion with a cut in Severtsov's subsidy.[19]

In the years 1869–1873 Severtsov resided in Moscow, St. Petersburg, and Petrovsk, where he digested the material gathered in central Asia, compared it with museum specimens, and wrote three important works. The first, *Travels in the Turkestan Region and Investigations of the Mountain System of Tian Shan* (1872), clearly reflected the military context of Severtsov's studies: its descriptions of the geography, flora, and fauna of Western Turkestan are interspersed with vivid accounts of battles between the tsarist forces and native peoples, and among the indigenous population itself.[20] The second, *The Vertical and Horizontal Distribution of Turkestan Animals* (1873), was a pioneering study in zoogeography and ecology; it was translated into English, French, and German and brought Severtsov international renown.[21] The third, "Arkary (mountain rams)" (1873), drew on material from his two books to analyze the evolution of one life form in light of its geographical distribution.[22]

In 1875, between expeditions to Amu Darya and Pamir, he again traveled to Europe, where he received a gold medal from the International Geographical Congress in France and visited Charles Darwin at Down. In a letter to Kaufmann, Severtsov claimed that Darwin had approved his pet project—a comprehensive treatise on evolution. This, he admitted, "has occupied me for some time, between other tasks."[23] By comparing his more than ten thousand central Asian specimens with the parent forms in other countries, he sought "to investigate and determine the laws of the variation of species and their successive development through the direct observation of birds."[24]

While in St. Petersburg for the Sixth Congress of Russian Naturalists and Physicians (1879), Severtsov heard Kessler's speech about mutual aid and commented sympathetically on it. He also met N. Ia. Danilevskii, who was soon to become Russia's most famous anti-Darwinist. The two engaged in a heated, all-night debate—Severtsov smoking thick *papirosy* and Danilevskii a long cherry pipe—that, by one account, "left an indelible memory upon the witnesses to this scientific contest."[25]

Among Severtsov's friends in these years was the president of the Imperial Russian Geographical Society, P. P. Tian-Shanskii. Tian-Shanskii's son later described his father's frequent guest:

> N. A. Severtsov was very homely; with his large size and massive stooping figure his appearance was frightening to children. This impression was heightened by the indelible lifelong traces of his skirmish with the Kokands, which he survived due only to his manliness and knightly nature: his face was scarred and one ear was slashed. Not only did he not conceal these defects, he readily exhibited his cloven ear to interested parties.... N. A. Severtsov's movements were very slow. He spoke unusually slowly and in a drawling manner, like the beat of ancient clocks, with a peculiar guttural sound, so everyone who told stories about him (and such stories and anecdotes were plentiful among those who had spent some time in Turkestan) attempted to reproduce his unique speech. I understood his style of appearance only when I saw an old Uzbek in 1888 at a bazaar in Tashkent; he and N. A. resembled each other like two drops of rain.... Clearly central Asia, with which he had become so intimate, had placed its stamp upon him. In his trips to the capital (which usually took place in the winter: in the summer N. A. worked on expeditions) Severtsov walked along the streets of St. Petersburg in an enormous, distinctive Siberian fur coat, which sometimes concealed the military uniform awarded him by the Governor General of Turkestan for his part in many military campaigns. Together with his typical scholarly appearance, his glasses and long hair, Severtsov had at these times an entirely improbable look and attracted considerable attention. This casual attitude toward his military uniform ... underlined his independent cast of mind.[26]

Casual acquaintances often thought Severtsov rude and affected. He could be a disconcerting conversationalist, sometimes responding to a remark long after the subject had passed, and then in words unfamiliar to urban intellectuals or in totally inappropriate stentorian interjections.[27]

An indefatigable worker, he was intellectually cautious to a fault and published relatively little after 1873. Kropotkin later recalled:

> A great zoologist, a gifted geographer, and one of the most intelligent men I ever came across, he, like so many Russians, disliked writing. When he had made an oral communication at a meeting of the [Imperial Russian Geographical] society, he could not be induced to write anything beyond revising the reports of his communication, so that all that has been published over his signature is very far from doing full justice to the real value of the observations and generalizations he had made.[28]

Severtsov never completed his comprehensive treatise on the evolution of birds nor any rounded statement of his mature views on the factors of evolution. In 1885 he was thrown from his horse through the ice of the Don River and died of shock shortly thereafter. He left behind a massive collection of notes and unfinished works—and one protégé, M. A. Menzbir.[29] Menzbir edited Severtsov's unfinished work on eagles, adopted his zoogeographical approach to evolutionary theory, and, as a one-man "Severtsov school," became an important defender of Darwinian orthodoxy in the next generation.[30]

Ecology and Evolution, 1855–1858

In his master's thesis Severtsov addressed "the periodic phenomena of animal life, their mutual connections and dependence on external conditions."[31] Like his mentor Rul'e, he thought cyclical activities such as migration, mating, and moulting reflected "determined relationships" between the organism and the environment and "no less determined life relations to other species of animals and to the other individuals of its own species."[32] Thus, periodic phenomena were governed by a multiplicity of inorganic and organic causes. Other theorists might attribute migration, for example, to the direct influence of cooler weather, but Rul'e and his student looked instead to interrelated changes in an entire complex of conditions including climate, food supplies, and defensive needs.[33]

For Severtsov, Russia, especially Voronezh, afforded ideal conditions for the investigation of these law-governed relationships. Here they were displayed under a great variety of physical circumstances, including "sharp extremes" of climate. Here too nature was still "pristine, . . . unaltered (except in a few provinces) by centuries of labor by a dense population."[34]

In light of his later reaction to Darwin's struggle for existence, two aspects of Severtsov's master's thesis are especially interesting: (1) his analysis of the influence of changing external conditions on relations between organisms and (2) his attention to differences between individuals of the same species.

Severtsov observed that the animals of Voronezh province differed greatly in their degree of sociability. They were "solitary or social, constantly social or social for a specific goal."[35] Sociability was conditioned by a variety of environmental conditions but was most sensitive to an animal's conditions of nutrition.[36] Examining the relationship between sociability and food supply among out-migrating birds, he concluded that (1) food consisting exclusively or primarily of vertebrates conditions a solitary and gradual out-flight, regardless of other circumstances; (2) food consisting of plants or invertebrates conditions "mixed sociability," depending on the presence or absence of suitable places for a flock to collect; (3) if all other conditions are equal, the species that eats more grains and grasses will be more sociable; and (4) periodic changes in food are accompanied by corresponding changes in sociability.[37]

The determining influence of the conditions of nutrition on sociability was especially dramatic among white-tailed eagles. When hunting, eagles sought their prey in solitary flight, spaced several versts apart (about two miles). When a lone eagle spotted a quarry, it cried out, alerting the others, who joined it in pursuit. "In this manner, the eagles can surveil a space of 40 versts length and width, maybe more." Severtsov once managed to get within 150 steps of some feeding whitetails before he was spotted. As they rose in flight, he noticed that the oldest ones flew with great difficulty—and thought this was because they were sated. The whitetails "eat in order of age, from older to younger."[38]

> If we recall that whitetails are birds that generally do not live in a society, doing so only with a specific goal, to eat, than we must conclude that their intermittent sociability arises from a series of instinctive actions and demonstrates a highly developed intelligence, which puts them on a much higher level than a majority of birds and on the same level as the wolf, which is also intermittently social, forming herds only with a goal.[39]

The whitetails hunted a wide variety of prey, Severtsov added, and so had become highly adaptive, resourceful creatures.

His close attention to the fluid relations among organisms was especially evident in his analysis of migration. For Severtsov as for Rul'e, migration resulted from external influences on the relations among different organisms. In the fall temperatures dropped and birds began to migrate—yet the relationship between the two events was indirect:

> If we keep in mind the simultaneity of the autumnal hibernation of reptiles, insects and certain beasts, the withering of plants, . . . and the fact that fish remain deeper in the water and come to the surface less frequently as the water grows colder, then it becomes very probable that the gradual, solitary out-flight, and even the outmigration in flocks, is conditioned only indirectly by the temperature, while its direct cause is the diminution of the food supply.[40]

Like his friend Kropotkin, Severtsov observed a great deal of cooperation in nature, yet his interest was captured by something that struck Kropotkin as insignificant: the existence of marked differences between individuals of the same species. He was intrigued, for instance, by the fact that with the onset of cold weather some birds migrated before others. "Probably, in each of the species mentioned above, the sensitivity of various individuals to cold is not identical, but subject to personal differences."[41] Similarly, "the time when moulting begins varies not only among various species but also among various individuals of the same species."[42]

These individual differences were central to his reflections about evolution in several works of the 1850s. Here he commented on the phylogenetic links between species, genera, and orders, and referred to the transformist views of Goethe, Geoffroy Saint-Hilaire, Oken, and Lamarck. He informed his readers that nuances in environmental conditions were reflected in the organisms that inhabited them, giving the animal kingdom the appearance "not of a chain, but more of a tree."[43]

In a lengthy unpublished manuscript on the zoogeography and classification of cats that he used as the basis for an oral presentation to Russia's Academy of Sciences in 1857, Severtsov summarized the zoogeographical approach to evolution that would guide him throughout his career. "Nobody has ever observed the process of species formation, and this is impossible, since it lasts thousands of years. But can one not see in variations a stage of this process? The facts make this possible."[44] In "Notice sur la classification multisériale des félidés, et les études de zoologie générale qui s'y rattachent" (1857–1858), he commented similarly that the mutability of species could be demonstrated "in a positive manner" by analyzing patterns of variation among contemporary forms. Characteristics that constituted only exceptional, individual traits in some populations defined species or genera in others. Zoogeographical studies of these characteristics could demonstrate their transformation from personal idiosyncracies within a single species into lines of demarcation between distinct species and genera.

In "Notice sur la classification" Severtsov demonstrated that a parallel series of individual idiosyncracies existed among lions, tigers, and other members of the cat family.[45] Localized and inconsistent within some races, these same traits constituted stable racial characteristics in others. For example, in some regions "brown- and black-bottomed jaguars are individual and exceptional variations; in others they form a race distinct from other forms. One sees . . . [here] that a local influence produces a tendency for individual characteristics to become permanent, to become racial characteristics, in a manner that differs from species to species."[46]

This fluid relationship between morphological characteristics and racial distinctions reflected the historical process of phylogenetic change. For instance, "the European lynx presents us with individual varieties at the very moment of their transformation, no longer into local races, but directly into species, each of which already manifests local variations."[47] So individual variations provided "the point of departure for races, and produce them upon becoming hereditary."

Severtsov summarized his evolutionary theory in a definition of "species":

> A species is an organic ensemble or a collectively organized being, like an individual. It lives in a series of generations, just as the individual loses and renews the cells and fibers that compose its body. It has its ages, its successive phases of development; it is modified in its entirety, as was maintained by Et. Geoffroy Saint-Hilaire. It generates derivative species which are transformed from initial, individual varieties into stable races, but which are linked by individuals with intermediate characteristics. They subsequently become completely separate. Depending on the local conditions of existence . . . the original type either continues to exist alongside the derivative forms or disappears; but these derived species always maintain in their characteristics a trace of a common descent, resembling one another more than they do the neighboring forms which constitute the genera and which were perhaps produced in the same manner.[48]

What was the source of these variant forms? Some speculated that they arose "by a principle of evolution, of modification, inherent to the organism" and largely independent of external influences. Others held that variations were the direct product of environmental influences. Despite misgivings, Severtsov provisionally supported the latter view because it was "more intelligible" and more amenable to "positive verification by observation."[49] He associated himself, then, with Lamarck's theory, "as modified by Et. Geoffroy Saint-Hilaire," that "under the influence of the environment the species type is modified either superficially or in its essential characteristics to the point of being transformed into another species over the generations."[50] These environmental circumstances produced variations by both direct and indirect means. Their direct influence altered those characteristics that varied most frequently. For example, the chemical effect of the sun's rays altered feather type and hair cover. More stable characteristics, such as the nature of the respiratory system, were transformed only indirectly and slowly as a changing environment altered their function, which led in turn to a change in form.

Darwin and the Struggle for Existence

Three years after reading the *Origin*, Severtsov wrote the following about evolutionary theory: "The best existing hypothesis is of course the Darwinist one, but it is inexact and subject to verification in too many places, and, for all the brilliance of its logical development, it is still far from having a sufficient factual basis."[51] He never changed this assessment.

Darwin's theory resonated with important aspects of Severtsov's own views. Like *Periodic Phenomena*, the *Origin* dealt with the multiplicity of intertwined relations between organisms and the environment and among organisms themselves. By 1873 Severtsov had accepted Darwin's "struggle for existence" as a useful metaphorical shorthand for these relations. Furthermore, Darwin's emphasis on intraspecific competition resonated with Severtsov's long-standing belief that differences between individuals of the same species held the key to evolution.

Nevertheless, as Menzbir later recalled, Severtsov's initial reaction to Darwin's theory was "restrained." "Perhaps longer than many others, he examined the new theory, not with distrust, but critically."[52] This was probably due both to Severtsov's tentative commitment to his own ideas about evolution and to his wariness about all generalizations. Darwin's theory struck him as powerful and plausible, but it relied too little on "positive knowledge" and too much on purely logical argumentation, especially the analogy with artificial selection.

He came, then, to accept the selection theory as a brilliant hypothesis requiring verification. It was "extremely useful and good as a philosophical theme for reducing all observed phenomena (in a generalized form) to the same denominator."[53] His research strategy remained essentially unchanged by his positive reaction to the *Origin*: he continued to seek zoogeographical evidence for the transformation of individual idiosyncracies into species characteristics. Yet he came increasingly to praise Darwin's theory, to analyze his own results by its light, and indeed to characterize his work as an attempt at "the factual verification of the Darwinist theory."[54]

Because Severtsov never completed a comprehensive statement of his mature views on evolution, the relative importance he assigned to natural selection and other evolutionary factors is unclear from published sources.[55] It is clear, however, that his examination of the struggle for existence in the field left him in complete agreement with Darwin's emphasis on the evolutionary consequences of intraspecific competition.

Severtsov addressed evolutionary theory most extensively in "Arkary (mountain rams)" (1873), an essay that he termed a "preliminary, immature" attempt to test Darwin's theory.[56] However impressive in appearance, he informed his readers, mountain rams led an uneventful existence. "They pasture year round, rut after females while in heat, and that is all." Precisely because their lives were dull, because they lived in easily defined environmental settings and had adapted to them through physical rather than mental developments, mountain rams presented the naturalist with an ideal subject "for the resolution of the most burning and capital questions of general biology."[57] The factors of evolution could be discerned by examining "the zoological influence and significance of external conditions, properly, of physical and biological ones, that is, of the conditions of the struggle for existence." Among the most important aspects of this struggle for existence was human activity, which had an influence on wild animals that was largely ignored by "Darwin and his school."[58]

The "classical land of the mountain ram" consisted of the ridges and hilly plateaus of High Asia, with their vast deserts, winds, terrible heat, and frost. Unlike the mountain goat, the ram did not live exclusively on cliffs. It led a nomadic existence near the pastures of the steppe and wandered about in small groups or in herds of eighty to one hundred. Its life was governed by two imperatives: to find food and to avoid the Khirghiz, who occupied the pastures with their domesticated stock.[59] Only when "the open steppe rages with irresistible snowstorms and forces the Kirghiz and their stock to seek more sheltered places" could the ram, "unafraid of the snowstorms," venture upon the vacated pasture.[60]

In order to ascertain their usual cause of death, Severtsov studied rams' remains and noticed that these often lay singly in the pasture land immediately proximate to a precipice. Judging from their degree of decay, he determined that a great number of males died in October, when rutting rams battled along the mountain precipices. The single skulls found under the precipices were those of weaker rams pushed over the edge by stronger ones. In rare instances the stronger male flung himself so powerfully against its adversary that both lost their footing, in which case "two skulls lie next to one another, no more than ten steps from each other; but more lie alone."[61]

Wolves sometimes took advantage of these battles; this served mainly to intensify indirect competition among rams since predators were unlikely to subdue "the stronger, more sensitive and dexterous" among them. So the fights among rams were useful to the species and constituted

> a simple but effective means of the natural selection of the strongest and most adroit sirers of offspring, who transmit to posterity their powerful muscles, elastic legs and enormous horns—generally, qualities that make it possible to hide from an enemy, jump along the rocks and enjoy the most inaccessible mountain pastures.[62]

After describing the mountain rams' mode of life, Severtsov detailed the different species and their geographical distribution. Ranging throughout southern Europe, northern Africa, Asia, and the western region of North America, mountain rams almost always lived along grassy slopes or in the partly mountainous areas that rose from the steppe. The habitats of different species differed primarily in their altitude and climate.[63] This "singularity, simplicity and clarity of the relationship between the conditions of place, climate, food and the struggle for existence" eased the naturalist's task considerably.[64] Identifiable differences in the conditions of the struggle for existence in different locales could be used to explain the fact that "personal" characteristics among one zoogeographical group, genus, or species were transformed into specific or generic characteristics in another.

Most rams belonged to one of two genera: the *Ovis* and the *Musimon*. *Ovis* was further divided into four zoogeographical groups. The species of each shared definite morphological characteristics, particularly their size and the thickness and curvature of their horns. The Tian Shan group, which included *O. poli* and *O. karelini*, all had extended horns with a complete spiral. The rams of northern and eastern Asia, such as *O. argali*, had shorter, thicker horns, with the same complete spiral. Those to the south in the Himalayas, such as *O. nahoor*, had small horns with a short, incomplete, reverse spiral.[65]

The structure of the fourth group differed markedly from the other three. Found in noncontiguous areas in North America and polar Siberia, it included *O. canadensis*, *O. montana*, and *O. nivicola*. Characteristics that defined the other groups were mixed in this one. *O. nivicola*'s horns combined the characteristics of the Asian and Himalayan types. The spiral of *O. montana*'s horns was intermediate to those of the genera *Ovis* and *Musimon*. Describing the end spans of the horns of another such form, Severtsov noted that

> its direction varies among various individuals, in one forward and out, in another directly forward, in still another forward and in; in the first case this span is longest, in the last it is shortest; I have seen photographs portraying a complete series of transitions between the extreme directions of the horned ends, so this characteristic, which is generic among every *Ovis* of the Old World (the ends of the horns turning forward), distinguishing this genus from every *Musimon* (the ends of the horns turning inward) becomes [merely] a personal trait [that varies from individual to individual] in the *O. Canadensis*.[66]

A similar pattern emerged with regard to the shape and thickness of horns.[67]

"To the reader the tips of horns and the differences in their direction will probably seem trivial," Severtsov continued, but they illustrated a critical fact: a characteristic that varied widely among individuals of one species, or among the species of one zoogeographical zone, became a stable, defining feature of species and genera in other zones. Thus, the relationship of a morphological characteristic to the definition of a species or genus was variable, it was "not unconditional, not primordial."

This had great significance for evolutionary theory:

> We will see further that these personal characteristics acquire generic significance in connection with differences in the conditions of the struggle for existence of wild sheep on the two mainlands. This is one of the direct proofs of the correctness of Darwin's theory—one not dependent upon an analogy between the characteristics of wild species and domestic strains—of which there are almost none in his well-known book on the origin of species, although these are not especially rare in nature.[68]

He developed this point further in a footnote:

> In Darwin's other works, also, I did not find examples of a specific or even a generic characteristic that was constant among some wild animals but was merely an inconsistent modification or personal trait among other, closely related wild forms. This I myself observed, for example, among eagles, thrushes, etc. Darwin provides no direct observations of the variability of wild animals and its relation to the formation of species; he only casually and in general expressions recalls disputed species of animals and plants. . . . I hasten to add that Darwin is correct: these direct observations are important for the critical verification of the great theory of natural selection, which he established so brilliantly.[69]

For Severtsov the wild rams of North America and polar Siberia represented the parent forms of those living in the other three zoogeographical regions. The areas in which these original forms lived differed from the other three in one essential feature: the virtual absence of domesticated breeds. He concluded that the evolution of Asian and southern European rams had been "decisively conditioned by the action of natural selection in the struggle for existence with domestic livestock, primarily with domestic sheep."[70]

> The discovery of such intermediate forms in America and polar Siberia, where [the forms between which they are intermediate] do not exist, and their absence in Asia, where the forms to which they are intermediate do exist, is explicable only by Darwin's theory, according to which these transitional forms are essentially the most ancient and fundamental; the American conditions of the struggle for existence facilitated their preservation while the Asian conditions facilitated the formation and separation from them of the very types between which they are now intermediate. Those offspring of these ancient, intermediate forms that did not change lessened in number with every generation until they disappeared.[71]

Characteristics that varied among the ancient forms were consistently present or absent in the forms that had evolved from them. This consolidation of traits had resulted from the differences in the conditions of the struggle for existence in different places. Severtsov conceded that the necessary details were still obscure. Yet two different aspects of the struggle for existence, each with a different morphological consequence, were clear. The conflict between rutting males clearly favored the development of large, quick, big-horned species.[72] In polar Siberia or extremely mountainous regions, however, undersized males would "gain the upper hand, because they were better adapted to the sparse pastures, which were insufficient for their larger competitors."[73]

Severtsov applied this same line of reasoning to the evolution of such related forms as goats, aurochs, antelopes, and domesticated stocks. The details need not concern us. We should note, however, his unproblematic use of the expression "struggle for existence" and his consistent emphasis on the role of intraspecific conflict. For instance, he attributed the separation of *Ovis* from *Capra* to the Ice Age, which "changed the conditions of the struggle for existence for each species, altering its competitors and facilitating the formation of new species by natural selection and the extinction of the former ones." Expanded mountain snows and ice had greatly reduced the amount of pastureland available to the original forms, leading to "a significant intensification of the struggle for existence among them." This led to the divergence of characters, as "only forms sufficiently varied not to disturb one another could be spared; those intermediate between them gradually perished."[74]

Domesticated forms had been removed from this struggle for existence before becoming distinct species, and artificial selection had preserved characteristics that would have proven fatal under natural conditions. These creatures had, however, exercised a profound influence on the subsequent evolution of their untamed relatives:[75]

> Protected by man, they reproduced more rapidly and, despite their weakness and flabbiness, surpassed the wild ones in number and pushed them higher and higher into the mountains, ascending themselves to the summer pastures. The occupation

of the wild sheep's pastures by herds of domestic rams produced a struggle for existence that was even more intense than that between the mountain sheep and mountain goats during the Ice Age.[76]

Those that perished most frequently in this intensified conflict were "first individuals and entire wild strains identical with domesticated ones, and then those closest in their qualities and characteristics to them." This resulted in the divergence between wild and domesticated types.[77]

Hunting further intensified the struggle for existence, and here again the essential consequence of interspecific conflict was intensified intraspecific competition. Hunters killed weaker individuals, and in some regions their predations led to the death of even those with average physical characteristics.[78]

Severtsov concluded on his customary cautionary note. His attempt to apply Darwin's theory was "still very insufficiently developed, as is each exact partial application of the theory of natural selection." His conclusions, therefore, lacked a positive, scientific character and represented only a "preliminary, immature effort."[79]

In his subsequent works he sometimes emphasized the direct influence of the environment rather than natural selection, but he consistently used the term "struggle for existence" unproblematically to encompass the totality of circumstances against which an organism contended. Indeed, he considered the investigation of the particularities of this struggle in different locales to be the central task of zoological geography:

> The main thing is the all-sided investigation, from various points of view, of the composition of fauna, determining for each the particular conditions of the struggle for existence.... The presence of organic types that are peculiar and exclusive to a given region is the most visible sign of the presence in this region of conditions of the struggle for existence that are favorable to them, and that are not found in other regions.[80]

Severtsov, then, was one of the few Russian naturalists who accepted Darwin's approach to the struggle for existence, including his emphasis on intraspecific competition. Why, then, did Kropotkin include him among the "great Russian zoologists" who, unlike Western Darwinists, saw, not struggle, but "a great deal of mutual aid" within a species?[81] While in France Kropotkin had read the following summary of Severtsov's remarks about Kessler's speech "On the Law of Mutual Aid":

> N. A. Severtsov, in support of K. F. Kessler's views, indicated an interesting example from the life of falcons. One species of these birds, despite a [physical] organization that appears remarkably well adapted to external conditions, is much less widespread than other forms, due to the absence among them of a tendency toward mutual aid.[82]

For Kropotkin this passage meant that "cooperation, not competition, is the key to survival." For Severtsov, however, individuals within a species could both cooperate and compete simultaneously; he wrote outside of the ideological context that encouraged Kropotkin, Kessler, Beketov, and so many other Russian naturalists to see cooperation and competition as opposing processes. Severtsov was simply observing, then, that mutual aid was widespread and advantageous to a species; it remained but one aspect of a struggle for existence that generated evolutionary change through intraspecific conflict.

His preoccupation with the differences between individuals of the same species, his high regard for Darwin, and his cautiousness about theoretical generalizations were evident in his final publication, "Etudes sur les variations d'âge des Aquilines paléarctiques et leur valeur taxonomique" (1885). Completed in the months before his death and prepared for press by Menzbir, this product of thirty years of research remained true to Severtsov's lifelong research strategy. In it he examined individual variations among eagles and their fluid relationship to species and genera. He praised Darwin fulsomely but noted that further studies of individual variations remained "indispensable for a positive solution to the great problem of the formation of species by natural selection."[83]

K. A. Timiriazev

> I have systematically avoided the unhappy expression "struggle for existence," which enemies of Darwinism exploit so unceremoniously.[84] (K. A. Timiriazev, 1901)

Kliment Arkad'evich Timiriazev (1843–1920) was the quintessential "man of the sixties." For him, science and democracy were allied in the struggle against despotism and suffering, and so he proudly described his life goals as "to work for science and to write for the people."[85] He was, indeed, both a painstaking laboratory scientist, who identified proudly with the Academician of Lagado lampooned in *Gulliver's Travels* for spending eight years staring at a cucumber in a phial, and a prolific propagandist, for whom philosophical materialism, Comtean positivism, the law of the conservation of energy, and especially Darwinism provided a unified, progressive worldview.[86] Timiriazev was Darwin's "Russian Bulldog," and his reaction to the Darwin-Malthus connection provides both an exception to the rule in Russia and a dramatic illustration of it.

He was born in St. Petersburg, a stone's throw from the statue of the Bronze Horseman overlooking the Neva River. Like so many budding scientists of his generation, he was the son of well-educated gentry who had fallen on hard times. His father had worked his way up the civil service to become director of customs but was dismissed in 1858 because of his leftist views. Kliment and his elder brother, Dmitrii (later a famous agricultural and factory statistician), were tutored at home, where they learned French and English and absorbed tales of the French Revolution and Russia's Decembrist uprising from their parents.

He was, then, ideologically well prepared to participate in the student demonstrations that rocked St. Petersburg University shortly after his matriculation in 1860. These protests against the limitations of the emancipation decree of 1861 led authorities to close the university. Offending students were readmitted the following year only if they signed a pledge of good behavior. Although "science was everything to me," Timiriazev refused and so was admitted only a year later as an "external student."[87] His first two publications, "Garibaldi in Caprera" (1862) and

"Hunger in Lancashire" (1863), expressed his progressive views and earned much-needed money.

At St. Petersburg University he studied botany with Beketov, chemistry with Mendeleyev, and zoology with S. S. Kutorga. The first two became his lifelong friends. The third introduced him to the *Origin* in a lecture of 1860; Timiriazev later recalled Kutorga's comment that "it is a new book, but a good one."[88] An immediate convert to Darwin's theory, Timiriazev presented it to Beketov's student study circle and became one of Darwin's first Russian expositors with his "Darwin's Book, Its Critics and Commentators" (1864).[89]

After graduation in 1866 he joined Mendeleyev's experimental agricultural station in Simbirsk, where he worked on improving fertilizers and agricultural techniques. Two years later, at Beketov's suggestion, he departed for western Europe, where he worked with physiologist Claude Bernard, chemists Marcelin Berthelot, Robert Bunsen, and Jean Boussingault; physicists Gustav Kirkhoff and Hermann von Helmholtz; and botanist Wilhelm Hofmeister. Timiriazev was especially taken with Berthelot, whom he praised as the "Lavoisier of the nineteenth century" for his progressive political views and synthetic approach to scientific questions. In these years Timiriazev drew on European resources to refine techniques for the spectral analysis of plants in order to analyze the process later known as photosynthesis.

On his return to Russia in 1870, he joined the faculty of the Petrovsk Academy of Agriculture and Forestry. There he founded Russia's first experimental station for the study of nutritive processes in plants, delivered courses on plant morphology, physiology, and systematics, and supervised the practical work of academy students. He was appointed a nonstaff dozent in the Department of Plant Anatomy and Physiology at Moscow University in 1872, and five years later became professor and chair of the department.

His pedagogical activities were curtailed by the political repression of the 1890s. In 1892 the authorities closed the Petrovsk Academy in response to the political "unreliability" of its students and faculty. Six years later Timiriazev was relieved of his professorship at Moscow University, ostensibly because, having served for thirty years, he now qualified for a pension. In 1902 he was stripped of his right to lecture at the university and remained only as head of the botanical cabinet. Nine years later he was among 125 scholars who resigned or were expelled from Moscow University in a conflict arising from the firing of a popular progressive professor and police repression of the resultant student protests.

An internationally acclaimed scientist, Timiriazev was invited in 1903 to give the Croonian Lecture to the Royal Society of London, to which he was elected in 1911. He also received honorary doctorates from Cambridge, Geneva, and other western European universities. He joined Russia's delegation to Cambridge University's 1909 celebration of the centenary of Darwin's birth and the fiftieth anniversary of the publication of the *Origin*. When he celebrated his own seventieth birthday in 1913, many European scholars, including Francis Darwin, traveled to Russia to honor him.

Throughout his career Timiriazev was a vocal critic of tsarism and tsarist ideology, denouncing police raids on the universities, defending radical students and faculty, and, in 1889, joining a student strike in honor of the recently deceased radical theoretician N. G. Chernyshevskii. After the revolution of 1905 he corresponded with Maxim Gorky and began to study Marxism. He denounced the imperialist nature of World War I and voted for Bolshevik candidates to the Moscow Duma and the Constituent Assembly in 1917. An enthusiastic supporter of the Soviet government, he served as a deputy to the Moscow soviet and sent a copy of his collection *Science and Democracy* to Lenin, "in consideration of the pleasure of being his contemporary and a witness to his glorious activity."[90]

Timiriazev's years of teaching produced many leading Russian academics of the late nineteenth and early twentieth centuries, including F. N. Krasheninnikov (Moscow University), V. I. Palladin (Kharkhov, Warsaw, and St. Petersburg universities), D. N. Prianishnikov (Petrovsk Academy of Agriculture and Forestry, Moscow University), V. I. Butkevich (Kharkhov University, Moscow Academy of Agriculture), L. A. Ivanov (Kharkhov Veterinary Institute, Moscow Academy of Agriculture), and P. A. Kossovich (St. Petersburg Institute of Forestry).

His influence, however, extended far beyond his formal students to the broad public that read his essays on science for over fifty years. These covered a wide range of subjects in the history of science but, like his scientific work, revolved largely around two lifelong passions: the plant's production of organic matter and Darwin's theory of evolution.

When Timiriazev died in 1920, thousands followed his cortege through the streets of Moscow. He was quickly enshrined as a hero of Russian and Soviet science.

Chlorophyll, the Conservation of Energy, and Darwin's Theory

For over forty years Timiriazev's research in plant physiology followed the program he enunciated in 1868:

> To study the chemical and physical conditions of [the formation of organic matter in plants]; to determine the components of the sun rays participating directly or indirectly in this process; to trace their subsequent fate inside the plant until their destruction, i.e., until their transformation into internal processes; to determine the correlation between the acting force and the work accomplished.[91]

Underlying these investigations was a philosophical mission: the defense and elaboration of the materialist worldview. In 1844 the English physicist John William Draper, best known today for his *History of the Conflict between Religion and Science* (1874), had concluded that the decomposition of carbonic acid in plants occurred most intensely where they absorbed the yellow rays of the sun. This was puzzling since the yellow rays contained less energy, and were absorbed less readily, than the red rays. Over the next twenty-five years experimental plant physiology came into its own, especially in Julius von Sachs's laboratory in Würzburg. One of Sachs's students, Wilhelm Pfeffer, sought to explain Draper's finding by denying the connection between solar energy and the plant's production of food.

For Timiriazev, Pfeffer's conclusion had unacceptable philosophical consequences. It suggested that the law of the conservation of energy did not apply to the production of food in plants, and thus that this pivotal natural process was dependent on a vital force. In decades of experimental work Timiriazev sought to substantiate the assertion by Robert Meyer, a founder of the law of the conservation of energy, that the plant consumed sunlight and transformed it into chemical forces that fueled photosynthesis. In his master's thesis, "The Spectral Analysis of Chlorophyll" (1871), and his doctoral essay, "On the Assimilation of Light by Plants" (1875), he argued that it was the red, not the yellow, rays that were most completely absorbed by chlorophyll and that were associated with the intensive accomplishment of photosynthesis. Chlorophyll was a sensitizer that, as Meyer had argued, absorbed and transmitted solar energy, leading to the formation of organic substances in the leaf. Timiriazev developed this perspective in numerous subsequent studies, including "The Plant as a Source of Energy" (1875), *The Life of the Plant* (1878), "The Quantitative Analysis of Chlorophyll" (1879), "The Amount of Useful Work Performed by the Green Leaf" (1882), "Solar Energy and Chlorophyll" (1883), "The Dependence of Photo-Chemical Phenomena on the Amplitude of the Light Wave" (1884), and "The Dependence of the Plants' Assimilation of Light on Light Tension" (1890). In his collection *Sun, Life and Chlorophyll* (1923), Timiriazev lauded Meyer and Helmholtz, and termed his own laboratory work a "comprehensive experimental response" to the demands made upon plant physiology by their discovery of the law of the conservation of energy.[92]

Timiriazev argued that chlorophyll's affinity for the red rays was explicable by the selection theory. "The manufacture of this curious substance [chlorophyll] by the plant economy," he wrote in 1877, "is a most striking example of the adaptation of organisms to environmental conditions." If the need to absorb a maximum amount of light were the only factor in a plant's coloration, he argued, all plants would be black. But black plants would be overheated by rays, such as yellow ones, with little energy potential.[93] Chlorophyll must have originally emerged among primitive algae, where it was still one of a great variety of pigments. Over time "it emerged victorious in the struggle for existence and then conquered the dry land."[94] So the green color of the plant was not, as even the hyperselectionist Alfred Russel Wallace had assumed, merely an accidental quality, like the color of minerals.[95] It was, rather, an evolutionary response to environmental demands and provided "the key to understanding the major, cosmic role of the plant in nature."[96]

His explanation of this "cosmic role" in "The Conclusions of a Century of Plant Physiology" (1901) provides a good example of Timiriazev's style as a publicist:

> The green leaf or, to be more precise, the microscopic green grain of chlorophyll, is the focus, the point in the world to which solar energy flows on one side while all the manifestations of life on earth take their source on the other side. The plant is the intermediary between sky and earth. It is a real Prometheus, stealing fire from heaven. The sunray stolen by it burns both in the flickering light of a wood splinter and in the blinding flash of an electric bulb; it sets in motion the mammoth flywheel of the gigantic steam engine, the painter's brush, the poet's pen.[97]

Timiriazev's second scientific passion was the explanation and defense of Darwin's theory. His "Darwin's Book, Its Critics and Commentators" (1864) was but the first in a series of essays and books, including "Darwin as a Scholarly Type" (1878), *Charles Darwin and His Doctrine* (1883), "Factors of Organic Evolution"

(1890), *The Historical Method in Biology* (1892-1895, 1922), and "The Significance of the Revolution in Natural Science Accomplished by Darwin" (1896). When Danilevskii's *Darwinism* appeared in 1885, and Strakhov touted it as the definitive refutation of Darwin's theory, Timiriazev responded in several polemical articles, including "Has Darwinism Been Refuted?" (1887) and "The Impotent Fury of the Anti-Darwinist" (1889).[98]

For Timiriazev, Darwin was biology's Newton. Earlier evolutionists such as Lamarck deserved credit for their perspicacity, but only Darwin had discovered the "real key" to the fit between an organism and its environment. The essence of Darwin's contribution was as follows:

> Variations, called forth by the blind play of physical forces, ... can incline either to the good or ill of the organisms, but that as a consequence of a historical process, which he metaphorically called the *struggle for existence* and *natural selection*, each useful deviation, each improvement, will be preserved while each harmful and useless one is suppressed, eliminated. So, the organism is not adapted to the environment under its direct influence, as Darwin's predecessors proposed; rather, over an endless series of generations, everything that corresponds to the harmony of the organic world is accumulated, and everything that contradicts it is eliminated.[99]

The factors of evolution were basically three: variation, which resulted from environmental influences, but which was usually not preadaptive; heredity, which accumulated variations and led to increasing complexity; and natural selection, that combination of "unlimited productivity and implacable criticism" responsible for the improvement of organisms.[100] Natural selection was driven by the struggle for existence, which, in turn, was generated by the "inexorable Malthusian law."[101]

In Timiriazev's view, Darwin's brilliance had failed him on only two minor counts—first, his formulation of the theory of pangenesis, which he had wisely recanted, and second, his use of the expression, though not the concept, struggle for existence.

Use, Defense, and Abandonment of Darwin's Malthusian Metaphor

When propounding the selection theory in 1864, the twenty-one-year-old Timiriazev adhered closely to Darwin's presentation and rhetoric. The *Origin* demonstrated that nature was at war and that the victims were many: "Instead of the eternally clear, smiling nature to which we are accustomed ... there arises before our astonished eyes a terrible chaos, where living beings are entangled in cruel, mortal combat, and where each living thing enters life over the corpses of millions of its own kind."[102] Organisms multiplied much more rapidly than the resources required to sustain them—this was "the inexorable Malthusian law applied to the entire organic world"—and so there transpired a "constant process of extermination."[103] This process, to which Darwin referred metaphorically as the "struggle for existence," had four aspects: struggle among individuals of the same species, between different species, with direct enemies (predator-prey relationships), and with the physical conditions of life.[104]

Like Darwin, Timiriazev insisted that the organism's direct relationship to physical conditions was of only secondary importance in the struggle for existence. "Much more important" was the indirect influence of these conditions on "the competition of organisms among themselves." What prevented a plant in the center of its habitat, and well suited to its physical conditions, from doubling or quadrupling its population? "Clearly, only the competition with other organisms. The [physical] conditions of existence participate only indirectly here."[105]

Neither Darwin's reliance on Malthus nor his expression "struggle for existence" presented any discernible problem for Timiriazev in 1864, nor did it in 1878, when he sketched the selection theory in *The Life of the Plant*.[106]

Why did Timiriazev not dismiss Malthus with a sneer, as did so many of his compatriots? Let me suggest two possible explanations. First, Malthus's fundamental insight (at least, in relation to organisms without consciousness) dovetailed with Timiriazev's philosophical commitment to the law of the conservation of energy and with the research on photosynthesis that flowed from it. Timiriazev often observed that food production was dependent on solar energy, and that this, ultimately, was a limited resource. For instance, in "The Plant as a Source of Energy" (1875), he emphasized that agricultural production could be increased a hundredfold by artificial means, but added that this did not invalidate Malthus's law. At some point in the distant future, humans "will have exhausted the ability to gain more food from the sun":

> Man will then make no further demand of the soil or his art, for more fuel or more food—he will not be able to get any more, because the sun will not be able to give any more. Then the law of Malthus will manifest itself in all its ominous cogency; mankind will have to take account of the death-rate before reproducing itself, as has been already anxiously suggested by perspicacious economists. No extra mouths, in the literal sense of the word, will then find room at the banquet of Nature. Will mankind ever attain this limit? By what new processes of synthesis will Berthelots of the future benefit it? What new sun-machines will be furnished by future Mouchots and Ericssons? Who can tell? One thing is certain, that our planet will acquire then a very dismal aspect. When man shall have arrived at the utilization of all the energy of the sun instead of only part of it as we do at present, then, instead of the emerald green of our meadows and woods, our planet will be covered with the uniform mournful black surface of artificial light-absorbers. . . . We cannot help shuddering at the idea of what life will be like when the earth is transformed into a universal factory, with no possible escape into the open even on a holiday, even for a single hour![107]

In 1878 he again commented on the relationship between his work on photosynthesis and an unavoidable issue for mankind. "I have in mind the issue connected with the Malthusian doctrine—the issue of the limit to the earth's productivity, the limited quantity of organic matter which man is in a position to acquire from a certain square of earth with the help of a plant." Chemists would certainly devise artificial sources of foodstuffs, but these too would be limited by the only available source of energy, the sun.[108]

Timiriazev repeated this point in every edition of *The Historical Method in Biology* from the early 1890s until his death. Rejecting Malthusian social prescriptions, he nevertheless warned against "rosy dreams" that man could escape indefinitely from the "nightmare of the Malthusian doctrine." For the naturalist, he insisted, two aspects of Malthus's argument were "obvious and irrefutable": organisms tend to multiply in a geometrical progression, and the quantity of life on earth is limited by the availability of solar energy.[109]

This latter point suggests a second reason for Timiriazev's relative tolerance of the Darwin-Malthus connection. An admirer of Auguste Comte, he believed that scientific knowledge transcended the metaphysical speculation and political interests that often influenced it.[110] Malthus's correct insights, then, were ultimately separable from his reactionary motives. When Russia's radical theorists split over the Darwin-Malthus connection—some seeing it as an indication of the bourgeois nature of the selection theory and others as a historical, but not an immanent, logical relationship—Timiriazev fit comfortably in the second camp.

Timiriazev initially ignored Russian critics of the Darwin-Malthus connection, dismissing them as "windbags" intent upon discrediting scientific progress. However, when Tolstoy's character Levin attacked Darwin in *Anna Karenina*, Timiriazev conceded that the word "struggle" had also misled honest and intelligent people.[111] For Levin, Darwin's theory illustrated the tension between reason and morality. "Reason discovered the struggle for existence and the law requiring one to throttle everybody who obstructs the satisfaction of his desires," he observed. "Reason could not discover love for another because this is unreasonable."[112]

Clearly, Timiriazev rejoined, Levin had never actually read Darwin's work, which demonstrated that "as applied to humans the struggle for existence signifies not hatred and extermination but, on the contrary, love and protection." For social beings struggle was not solely individualistic. It was also "a struggle or competition between collected units—between families, tribes and races; in this struggle, or, more accurately, competition, success depends equally upon material energy, intellectual superiority, and moral qualities in relation to one's own kind."[113]

As we have seen in Chapter 2, Darwin's theory attracted considerable moral censure in the 1880s, most notably from the conservatives Strakhov and Danilevskii and the radical Chernyshevskii.[114] All three authors castigated Darwin for using the Malthusian progressions and implying that struggle among organisms was a source of progress. Timiriazev responded acerbically to conservative intellectuals, who, he claimed, misunderstood Darwin's theory entirely. "Darwinism" (meaning Darwin's ideas and those of his correct interpreters) did not hold that the struggle for existence was a universal moral law to which man was subordinate. Nor did it see victory as the result of "vulgar force" in direct combat. Quite to the contrary, "success is always on the side of the *higher* power," and so the selection theory provided a scientific basis for faith in the gradual moral and intellectual progress of mankind.[115]

Chernyshevskii's "Origin of the Theory of the Beneficence of the Struggle for Life" (1888) presented a more sensitive problem.[116] Timiriazev greatly admired the radical theorist and knew that his critique of Darwin's Malthusianism would carry great weight among progressive intellectuals. Yet he did not respond explicitly to Chernyshevskii's argument, explaining years later that it would have been unconscionable to do so while Chernyshevskii was muffled by the censor.[117]

His most telling reaction was a silent one. Beginning with *The Historical Method in Biology*—which originated as lectures in 1890-1891 and was first published in

1892-1895—the expression "struggle for existence" largely vanished from Timiriazev's explanations of the selection theory. He indicated several times in subsequent years that this omission rendered Darwin's theory more comprehensible and palatable to his Russian audience:

> I call the expression "struggle for existence" unfortunate since I am convinced that the polemic against Darwinism as an allegedly unethical doctrine relies mainly on mere words and has no content. Proof that this expression is unnecessary is the fact that I succeeded [in 1890-1891] in reading an entire Darwinist course ... without once uttering the word *struggle* [1895].[118]
>
> Nothing, perhaps, has brought such harm to [Darwin's] doctrine as this metaphor, which he could have avoided had he foreseen the conclusions that would be drawn from it [1896].[119]
>
> I have systematically avoided the unhappy expression "struggle for existence" which the enemies of Darwinism exploit so unceremoniously [1901].[120]
>
> This metaphor is unnecessary to the doctrine of natural selection and has only given birth to countless misunderstandings [1907].[121]

Timiriazev further adapted *The Historical Method in Biology* to Russian sensibilities by emphasizing the distinction between Malthus's valuable insight into nature and his reactionary political doctrine. There was nothing pernicious about the observation that population tended to outstrip food supply; indeed, this truth had been first discovered by the great progressive, Benjamin Franklin. "One can be certain that had Darwin used the popular name of Franklin instead of the antipathetic (for many) name of Malthus he would have avoided certain acrimonious criticisms."[122] Malthus's political economic theory, on the other hand, had been wisely and "indignantly rejected by Russian economists."[123] In Malthus's haste to banish the poor from nature's banquet, he had begged two important questions: "How many courses do those sitting at the banquet get and would it not be more just, before depriving anybody of a place, to ensure as just a distribution of the available viands as possible? And then a second question arises: are all the viands that nature can provide available at the banquet?"[124] Until these questions were resolved through social reforms, "not on paper, of course, as Malthus has done," human conflict and suffering remained the result of human arrangements rather than natural law.

For Timiriazev it was ironic, but hardly ominous, that the same law whose discovery inhibited social justice was responsible for progress in the natural world:

> If, in application to the conscious activity of man, this [Malthusian] doctrine served as a justification for the premature, passive subordination of moral indignation, the justification for the most soulless *inertia*, then this same principle, acting in the sphere of the unconscious factors of nature, is a necessary, mechanistic cause of the *progress* that characterizes the historical development of the organic world.[125]

This distinction became a staple of his subsequent essays. "Not the facts, but only the conclusions drawn from Malthus's book," he claimed, "are antipathetic."[126]

Darwin's theory itself, then, was innocent of Malthusian individualism, conservatism, and pessimism. "Neither Darwin nor any consistent Darwinist," Timiriazev insisted in 1901, "had ever extended the idea of the struggle for existence to "the cultured human of today."[127] In "From the Deed to the Word, From the Beast to Man" (1907) he decried the common tendency to blame Darwin for militarism and the decline of ethics. Humans had the ability to "paralyze" the organic law of the struggle for existence, to convert enemies rather than exterminate them, and to cooperate for the common good.

Concern about Darwin's reputation as a Malthusian probably was also responsible for Timiriazev's uncharacteristically tolerant reaction to Kropotkin's *Mutual Aid*. He usually responded swiftly and unforgivingly to any intellectual tendency, such as "Mendelism," which he feared might undermine or replace the selection theory.[128] Yet he limited himself to the tepid assessment that Kropotkin's book was "interesting" and found fault only with Kropotkin's comment that Kessler had been the first Russian to object to the use of the term "struggle for existence" with respect to man. Timiriazev insisted that *he* had been first, in the above-cited passage from "Darwin as a Scholarly Type."[129]

So by the 1890s Timiriazev had adapted his presentation of Darwin's theory to the sensibilities of his Russian audience by dropping the term "struggle for existence" and by separating Malthus's correct observations from his unconscionable political theory. He took a further step in later essays by interpreting the "struggle for existence" through another metaphor more congenial to Russians. In 1864 he had described the "cruel, mortal combat [in which] each living thing enters life over the corpses of millions of its own kind." By 1907 he was explaining that, for Darwinists, the struggle for existence was like "a drowning person struggling for life" against the sea. Unfortunately, some had "taken it incorrectly as a fight in which one party perishes."[130] (A more rigorous use of Timiriazev's metaphor, though one less likely to please a Russian audience, would have compared the struggle for existence to "drowning people struggling for life against the sea, with some better able to swim than others.")

This rhetorical emphasis on the struggle against physical conditions blended neatly with the political conclusions that Timiriazev thought most compatible with Darwinism. In a public lecture entitled "The Plant's Struggle with Drought" (1892), he dismissed a Russian jurist's argument that Darwin's popularization of the struggle for existence encouraged war and immorality. For Darwin the word "struggle"

> does not mean extermination of one's own kind, but only self-defense—the victory of life over the hostile forces of organic nature. And man, it seems, could confidently imitate such a struggle. If the energies wasted in mutual struggle, implicit or open, were concentrated harmoniously upon the bloodless battle with nature, if even a portion of the labor and knowledge wasted on the invention of weapons of extermination ... had been turned instead to the study and subordination of nature, then, of course, such calamities as drought and hunger would already belong to history.[131]

He elaborated this position in "Charles Darwin and Karl Marx" (1919), which celebrated the fundamental similarities between these two great intellectuals. Both had employed a historical approach to uncover the dependence of varied phenomena on underlying material conditions, and both had demonstrated that the intellectual and moral development of mankind rested on the social instinct and communality.[132]

Timiriazev summarized his sentiments about Darwin's metaphor, and the contradictory rhetorical strategy by which he defended it, in one sentence written in 1901: "Every rational, cultured human activity is part of the same struggle—the struggle against the struggle for existence."[133]

The Limitations of the Classical Tradition

We have dealt here with Severtsov and Timiriazev as exceptions to the rule. Yet they were also much more than that. Together with Menzbir and paleontologist V. O. Kovalevskii, they constituted an intellectually rich vein of classical Darwinian thought. Overpopulation and intraspecific conflict played an important part in Severtsov's analysis of the evolution of the mountain ram, Menzbir's treatises on birds, Kovalevskii's reconstruction of the history of ungulates, and Timiriazev's scientific and polemical essays. Severtsov, Kovalevskii, and Timiriazev were also bound to Darwin by fond memories of a pilgrimage to Down; in addition, Kovalevskii cherished his ongoing correspondence with the British naturalist.[134]

However impressive this foursome's intellectual achievements, they proved unable to defend Darwin's theory effectively against their compatriots' charges of Malthusianism. Two of them, Severtsov and Kovalevskii, played little part in the institutions, circles, and popular journals that defined the intellectual contours of the urban intelligentsia. Severtsov never held an academic post, and Kovalevskii, although widely respected in western European circles, managed to become a dozent at Moscow University only two years before his suicide. Neither offered any comment in their published work about the Darwin-Malthus relationship and its significance for evolutionary theory. Menzbir and Timiriazev, on the other hand, were both institutionally well positioned as professors at Moscow University. Yet Timiriazev's defense of Darwin's metaphor culminated in his abandonment of it, and Menzbir avoided the subject almost entirely in his writings before 1917. Their defensive posture was reflected clearly in their public reaction to Kropotkin's *Mutual Aid*; each responded, not with a vigorous refutation of Kropotkin's departure from Darwinian orthodoxy but, rather, with faint praise. With respect to differing estimations of the struggle for existence, Timiriazev and Menzbir practiced the politics of inclusion—they welcomed Kropotkin, like Beketov, Mechnikov, Kessler, and other like-minded thinkers, into the Darwinian camp.[135]

Thus, there was no "struggle over the struggle for existence" among Darwin's Russian admirers. Not, that is, until the emergence of a much different set of circumstances in the years after 1917.

Conclusion

He stops a metaphor like a suspected person in an enemy's country.[1] (Charles Lamb)

We have considered a most diverse group of Russian intellectuals. They were radicals, liberals, and monarchists; atheists, agnostics, and devout believers; political thinkers, novelists, and naturalists; laboratory scientists, museum curators, and field zoologists. For some, Charles Darwin was the greatest naturalist of the age, for others he was an important contributor to evolutionary theory, and for still others he was the author of a mistaken and pernicious doctrine.

We have seen that these same people, who diverged so radically in their backgrounds, viewpoints, and preoccupations, nonetheless had a common response to Darwin's "struggle for existence" and its Malthusian associations. This response was *national* in two important respects. First, it was pervasive: The thinkers I have discussed were central to the journals, circles, and life activities that defined the Russian intelligentsia. One could not take an interest in philosophy, political theory, or science—one could be neither a Westernizer nor a Slavophile—without confronting the issues raised by Darwin's metaphor. These issues found their way into Tolstoy's novel *Anna Karenina* and a collection of children's tales; into Chekhov's "On the Farmstead" and a description of Martian history.[2] Second, this response to Darwin transcended the reference group categories that have proven so useful in social histories of science. As we have seen, it was not limited to one or another political tendency—uniting the nihilist Nozhin, the populist Mikhailovskii, the liberal Bibikov, and the monarchist Pobedonostsev. Nor was this response specific to a discipline—Korzhinskii was a botanist, Bogdanov a zoologist, Kropotkin a geologist, Bekhterev a psychiatrist, and Skvortsev a hygienist. It was peculiar neither to a generation (Kessler was born in 1815, Mechnikov in 1845, Korzhinskii in 1861) nor to a particular type of scientific practice (Brandt was museum-based, Mechnikov worked in the laboratory, and Dokuchaev in the field). Critics of Darwin's metaphor were concentrated in St. Petersburg—as was the intelligentsia itself—but also included professors and students in Moscow, Dorpat, Kazan, Kharkov, Kiev, Odessa, and Tomsk. This response defies identification with a single teacher, patron, circle, or institution.

While these reference groups can not account for the Russian response itself, they clearly influenced its contours and the fine grain of individual reactions. Two institutions in the capital, St. Petersburg University and the St. Petersburg Society of Naturalists, seem especially important in this respect. As chairs of their respective university departments and sometime presidents of the society, Beketov and Kessler exercised considerable influence on scientific training and research. Both spoke frequently at the society, which became a stronghold of support for Kessler's theory of mutual aid, and both encouraged society-sponsored field studies of natural relations in different locales. Informal relations among the members of the society, who also included Bogdanov, Brandt, Dokuchaev, Kropotkin, Poliakov, and Tanfil'ev, no doubt encouraged skepticism about Darwin's metaphor. Indeed, the shift in Bogdanov's portrayal of intraspecific relations after 1873 may well have been influenced by his St. Petersburg contacts in the 1870s, especially his increasingly warm relationship with Kessler. The informal circles (*kruzhki*) that populated each urban center, and the networks that linked them, no doubt played a similar role. Mechnikov's personal contacts, for example, included members of the radical community (his brother Lev, Herzen, Bakunin, Nozhin), scientists (Beketov, Sechenov, the Kovalevskii brothers), and the author Tolstoy. It is likely that this group shared ideas about Darwin and the struggle for existence and that the similarities between Mechnikov's essay of 1863 and Nozhin's of 1866 owed something to their contact. Yet Mechnikov's close relationship with V. O. Kovalevskii and his rapport with Tolstoy also remind us of the obvious: personal contacts do not necessarily lead to agreement about substantive issues.

Differing philosophical and political viewpoints were, of course, evident in individual reactions to the struggle for existence. Alert to the weaknesses of the radical scientism of the 1860s, Lavrov and Mikhailovskii identified Malthusian elements in Darwin's theory and reformulated his struggle for existence to fashion an evolutionism compatible with their own populist views. Determined to preserve the scientific ethos of that same 1860s tradition, Tkachev, L. Mechnikov, and Plekhanov argued that the Darwin-Malthus connection was superficial and emphasized the qualitative difference between struggle in nature (accurately portrayed by the great scientist Darwin) and in society (tendentiously distorted by the corrupt ideologue Malthus). Nozhin, Tolstoy, Strakhov, and Danilevskii used the Darwin-Malthus connection to illustrate a more general point about science: for Nozhin the dangers of bourgeois ideology in the guise of natural history; for Danilevskii, Strakhov, and Tolstoy the cultural specificity of Darwin's theory and of all ostensibly objective knowledge.

The diverse reactions of leading scientists were related both to ideological tendencies held in common with the lay community and to other factors specific to their scientific practice. For example, debates about the division between forest and steppe led Korzhinskii and Tanfil'ev to consider the struggle for existence within a particular constellation of issues, including some specific to botanical geography (for example, what plant species can survive in a particular soil? what explains the geographical distribution of *Tussilago farfara*?) and some of much broader interest (for example, vitalism versus mechanism, inequality in nature and society). Both Korzhinskii's and Mechnikov's attitudes toward the struggle for existence changed over time with the changing focus of their scholarly work—for Korzhinskii from botanical geography to evolutionary theory, for Mechnikov from zoology to pathology. These changes confronted each with a different set of issues related to Darwin's metaphor, contributing to Mechnikov's growing appreciation of Darwin's theory and to Korzhinskii's ultimate rejection of it. The particularities of Timiriazev's scientific work and philosophy seem also to have influenced his thinking about the struggle for existence, predisposing him to accept the "positive" kernel of Malthus's theory.

Mutual aid theorists also exhibited a wide variety of views. The ichthyologist Kessler, impressed by the hazardous journeys of spawning fish, thought cooperation originated with family ties and the drive to reproduce; the ornithologist Bogdanov stressed the need for protection against predators and the elements. For the anarchist Kropotkin, natural communities were most often egalitarian; for the monarchist Brandt, they were benignly hierarchical.

Yet the very nature of this diversity underscores the fundamental unity of the Russian response: The great majority of these intellectuals responded negatively to the Darwin-Malthus connection and the metaphor "struggle for existence." Russians dismantled that metaphor, examined the relationship and relative importance of its component parts, and deemphasized or denied the role of those most closely identified with Malthus: overpopulation and intraspecific conflict. The national character of this response is also clear from its targets. Russians directed their arguments almost exclusively against non-Russian thinkers: Darwin, Wallace, and Malthus; unnamed European "Darwinists," "Social Darwinists," and "Malthusians"; the English bourgeoisie and the "British national type"; and Nietzsche and other unnamed advocates of "the right of the fist." We may wonder why Russians so rarely criticized their compatriots. Perhaps it was because the vast majority were in basic agreement—and the minority stood to gain little by challenging the consensus.

I suggest that this unifying national style flowed from the basic conditions of Russia's national life—from the very nature of its class structure and political traditions, and of its land and climate. Russia's political economy lacked a dynamic, pro-laissez faire bourgeoisie and was dominated by landowners and peasants. The leading political tendencies, monarchism and a socialist-oriented populism, shared a cooperative social ethos and a distaste for the competitive individualism widely associated with Malthus and Great Britain. Furthermore, Russia was an expansive, sparsely populated land with a swiftly changing and often severe climate. It is difficult to imagine a setting less consonant with Malthus's notion that organisms were pressed constantly into mutual conflict by population pressures on limited space and resources.

This combination of anti-Malthusian and non-Malthusian influences deprived Darwin's metaphor of commonsensical power and explanatory appeal, generating questions about the struggle for existence and predisposing Russians against accepting overpopulation and intraspecific conflict as important factors in evolution. They greeted Darwin warmly, like a long-awaited guest, but scrutinized his Malthusian metaphor coolly, "like a suspected person in an enemy's territory."

Was this Russian national style unique? In one sense it clearly was not. Yvette

Conry has identified anti-Malthusian tones in some discussions of Darwin in France, as has Paul Weindling in Germany. Furthermore, at the turn of the century the German biologist Ludwig Plate developed an extensive classification of the struggle for existence.[3] This should not surprise us. The Russian response, after all, followed, not from a mysterious "Russian soul," but from the confluence of two specific conditions, one socioeconomic, the other physicogeographical. Variants of these conditions must have influenced discourse in other countries as well.[4] Yet they seem to have combined with particular intensity and effect in Russia, framing the terms in which Russian scientists, as a group, discussed and developed Darwin's theory. The secondary literature suggests that criticism of Darwin's metaphor was not a central feature of scientific discourse in other countries, but this remains to be determined by comparative studies.[5]

In the absence of such studies, let me suggest some tentative comparisons with Great Britain. The Englishmen who simultaneously developed the selection theory, Darwin and Wallace, shared two experiences: a sympathetic reading of Malthus's *Essay on Population* and an important field experience in tropical rain forests. Most Russian evolutionists shared two experiences that were roughly opposite to these: an aversion to Malthus and life on a vast continental plain.

The fate of Malthus's ideas in Great Britain and Russia could hardly have differed more profoundly. In Great Britain Malthus exercised a "magisterial influence on public opinion," and for leading British evolutionists his *Essay on Population* was a "great work" (Darwin) that elaborated "incontrovertible" arguments (Hooker) "most minutely and truthfully" (Huxley).[6] In Russia Malthus's *Essay* did not even find a translator until 1868, by which time its author had been widely identified as the embodiment of western European, particularly British, vices. He was a "hack writer" (Tolstoy) whose "purely English doctrine" (Danilevskii) was a "morally repugnant" (Beketov) expression of the "secret desires of the wealth-producing classes" (Kropotkin).[7] For Darwin and Wallace, Malthus was a reputable source of knowledge. For most Russian intellectuals he was part of a troublesome paradox: What, as Tkachev put it, could "the great Darwin" have in common with "the pastor-thief"?[8]

Not only did Russians read Darwin's explicit references to the despised Malthus, they could see with a glance that Darwin's Malthusian bias had distorted his perception of nature. Darwin described an essentially plenitudinous nature in which organisms were packed tightly, wedgelike, into every available space, where a small advantage could bring prosperity to one form only at the expense of another. Yet Russian nature was a great, sparsely populated plain. Where were Darwin's wedges? Here population growth was most obviously checked by physical circumstances, and these were often so severe that one form's slight advantage over another could easily seem insignificant. A sudden blizzard or an intense drought appeared to obliterate entire populations of insects, birds, and cattle without regard for the differences among them.

The challenge posed to Malthusian thinking by Russia's land and climate is most striking, I think, in the comments of two Malthusians. Malthus himself was impressed by Russia's underpopulated vastness and its untapped wealth; he applauded tsarist efforts to increase the empire's population and found in Russia persuasive evidence that feudal social relations could impoverish even a spacious land rich in resources. Butovskii praised Malthus as the "Galileo of political economy" and considered his general principles unassailable, but he conceded that they were, of course, inapplicable to "our broad and expansive Russia."[9]

Similar non-Malthusian perceptions informed Russian comments about Darwin's theory. Danilevskii, who had traveled for decades throughout Russia, rejected Darwin's vision of a plenitudinous nature—he saw not wedges but "wide gaps [and] voids." Nature, he insisted, did not resemble a single fountain at which a multitude of organisms battled to drink but, rather, the shore of a long river, where any crowding at one spot merely necessitated a short trip to another. For Dokuchaev, Darwin's struggle for existence indeed obtained in tightly constricted areas, but certainly not "over stretches of thousands of versts of black soil." Even Mechnikov, a laboratory scientist without field experience, found in the Russian landscape an obvious refutation of Darwin's assertion that struggle resulted from overpopulation.[10]

Russians were familiar with other areas of the world, of course, and even the tsarist empire contained some tropical regions. Yet they tended to perceive the great continental expanse on which they lived, and which they were encouraged to study as the tsarist empire expanded inland, as paradigmatic of essential relations in nature. For example, in his famous volume on Siberia, Middendorf commented that this region's "very scarcity of a variety of animal forms facilitates a better understanding of the general laws of life." He contrasted this with the tropics, where a profusion of organic forms obscured the fundamental relationship between organisms and the physical environment and so prevented naturalists from "penetrating deeper into the subject."[11]

To what extent did tropical nature play a similarly paradigmatic role for leading British thinkers? I do not know, but certainly Middendorf's assessment of the relative benefits of studying Siberia and the tropics contrasted sharply with Wallace's. For Wallace it was tropical nature, with its abundance of life forms and relatively stable physical conditions, that provided the ideal setting to study the natural relations most important for evolution. Like Darwin, he was well aware that the struggle for existence had a different character in the tropics than in temperate and frigid zones. The tropics "must always have remained thronged with life, and have been unintermittingly subject to those complex influences of organism upon organism which seem the main agents in developing the greatest variety of forms and filling up every vacancy in nature." In the temperate and frigid regions, on the other hand, "a constant struggle against the vicissitudes and recurring severities of climate must always have restricted the range of effective animal variation."[12] He concluded that in tropical regions "evolution has had a fair chance," while elsewhere "it has had countless difficulties thrown in its way."[13] For Wallace, then, the natural relations central to evolution were those observable in the tropics. Those of other regions, such as Middendorf's Siberia, were of peripheral importance.

Compared with Wallace's, Darwin's tropical experience was paltry, a matter of months rather than years. Yet he wrote in 1876 of his *Beagle* voyage that "the glories of the vegetation of the Tropics rise before my mind at the present time more vividly than anything else."[14] Nor was this a merely retrospective impression.

There was an almost religious quality to the young Darwin's reaction to tropical nature. "No disciple of Mahomet ever looked to his seventh heaven, with greater zeal, than I do to those regions," he enthused in one letter of 1833.[15] He explained in his *Journal of Researches* that "the elegance of the grasses, the novelty of the parasitical plants, the beauty of the flowers, the glossy green of the foliage, but above all the general luxuriance of the vegetation, filled me with admiration."[16] The details of tropical life might fade from memory, he predicted, but they "will leave, like a tale heard in childhood, a picture full of indistinct, but beautiful figures." Here he proved remarkably prescient, for tropical nature survived in one of the more powerful metaphors in his *Origin of Species*. The young Darwin had described the dense biota of the tropics as an "entangled jungle"; in the *Origin* he would refer to all of nature metaphorically as an "entangled bank."[17]

Historians of science have only begun to illuminate the many factors that influence scientists' experience and interpretation of nature. I hope this volume has contributed something to that endeavor. I also hope it will encourage others to study the way metaphorical expressions mediate between scientific thought and the circumstances in which it develops. To what extent, and with what consequences, have metaphors in science originated in nation-, class-, race-, and gender-based ideologies; in disciplinary agendas and institutional settings; and in specific experiences in the laboratory or the field? What happens to these metaphors when they travel across such lines of demarcation? Examination of the varied reception of such expressions could reveal specific experiences and ideologies underlying scientific ideas.

A related task—one I have not assumed in this volume—is to study how the relationship between a metaphor and scientific discourse changes over time. For example, we know that the themes explored in the present volume retained a rich resonance in the years after the Bolshevik Revolution. They are evident in G. F. Gauze's *The Struggle for Existence* (1934), with its experimental verification of intraspecific conflict; in T. D. Lysenko's "creative Darwinism," with its condemnation of Darwin's "Malthusian error" and its denial of the evolutionary significance of overpopulation and intraspecific conflict; and in the attention devoted to the "problem of the struggle for existence" by Soviet historians and philosophers of biology. What was the relationship between pre- and post-revolutionary arguments? How were discussions of the struggle for existence altered by successive phases in the Soviet Union's political economy and ideological life, by the changing organization and composition of its scientific community, by international ties, and by the rise of experimental biology?[18]

Darwin employed many metaphors. It would be interesting to know more about their fate in various settings. It seems appropriate that the title essay of Robert Young's book about British thought, *Darwin's Metaphor*, deals with an expression that Russians found unproblematic, "natural selection." Wallace's article "Mr. Darwin's Metaphors Liable to Misconception" (1868) had the same focus. He mentioned the struggle for existence only as a self-evident truth, apprehension of which enabled one to grasp the meaning of the problematic "natural selection."[19] In a letter to Darwin in 1866, Wallace mentioned that the struggle for existence was a fact "which no one of your opponents, as far as I am aware, has denied or misunderstood." The term "natural selection," however, with its personification of the selecting agent, so confused the British audience that Wallace urged Darwin to drop it altogether.[20] By then, of course, it was too late. Darwin could only continue to assert in the *Origin*, as he had since 1861, that "every one knows what is meant and is implied by such metaphorical expressions."[21] Yet this was surely wishful thinking. The fate of his theory in Russia demonstrates that culturally specific metaphors can have a substantial impact on the reception and elaboration of scientific ideas.

NOTES

Introduction

1. As we shall see in Chapter 1, Darwin metaphorically characterized a broad range of relations between organisms and abiotic circumstances, and among organisms themselves, as "struggle." For example, his "struggle for existence" encompassed not only predator-prey and host-parasite relations, but also the dependence of a plant on moisture, and the relative success of two individuals living miles apart in securing resources and leaving progeny.

On the metaphorical dimensions of the "struggle for existence" see Gillian Beer, "'The Face of Nature': Anthropomorphic Elements in the Language of *The Origin of Species*," in L. J. Jordanova, ed., *Languages of Nature* (New Brunswick: Rutgers University Press, 1986), pp. 207–43; Howard Gruber, "The Evolving Systems Approach to Creative Scientific Work: Charles Darwin's Early Thought," in Thomas Nickles, ed., *Scientific Discovery: Case Studies* (Dordrecht and Boston: Reidel, 1978), pp. 113–30; Stanley Edgar Hyman, *The Tangled Bank: Darwin, Marx, and Freud as Imaginative Writers* (New York: Atheneum, 1962), pp. 9–78; and Edward Manier, *The Young Darwin and His Cultural Circle: A Study of Influences Which Helped Shape the Language and Logic of the First Drafts of the Theory of Natural Selection* (Dordrecht, Boston, and London: D. Reidel, 1978).

For examples of other metaphors in biology see Arthur Lovejoy, *The Great Chain of Being: A Study of the History of an Idea* (New York: Harper & Row, 1936); L. J. Jordanova, "Naturalizing the Family: Literature and the BioMedical Sciences in the Late Eighteenth Century," in Jordanova, *Languages of Nature*, pp. 86–116; Robert Young, "Darwin's Metaphor: Does Nature Select?" in Young, *Darwin's Metaphor* (Cambridge: Cambridge University Press, 1985), pp. 79–125; Nancy Stepan, "Race and Gender: The Role of Analogy in Science," *ISIS* 77, 287 (1986): 261–77; and Emily Martin, *The Woman in the Body: A Cultural Analysis of Reproduction* (Boston: Beacon Press, 1987).

On the nature of metaphor see especially Max Black, *Models and Metaphors* (Ithaca, N.Y.: Cornell University Press, 1962) and "More About Metaphor," in Andrew Ortony, ed., *Metaphor and Thought* (Cambridge: Cambridge University Press, 1979), pp. 19–43; Richard Boyd, "Metaphor and Theory Change: What is 'Metaphor' a Metaphor for?," in Ortony, *Metaphor and Thought*, pp. 356–408; Mary B. Hesse, *Models and Analogies in Science* (Notre Dame, Ind.: University of Notre Dame Press, 1966); and C. M. Turbayne, *The Myth of Metaphor* (Columbia: University of South Carolina Press, 1979).

2. For Kropotkin's and Timiriazev's comments about Russian anti-Malthusianism, see chapters 7 and 8 of the present volume. See also M. A. Antonovich, *Charl'z Darvin i ego teoriia* (St. Petersburg, 1896), pp. 90–104. For Lysenkoist observations about the anti-Malthusianism of nineteenth century Russian social thinkers and biologists, see, for example, V. V. Makhova, P. V. Makarov, and K. Iu. Kostriukova, *Obshchaia biologiia* (Moscow: Medgiz, 1950), pp. 212, 230, 231, 241; E. A. Veselov, *Darvinizm* (Moscow: Uchebno-Pedagogicheskoe Izdatel'stvo, 1955), pp. 199–203; B.

E. Raikov, *Predshestvenniki Darvina v Rossii* (Leningrad: Uchebno-Pedagogicheskoe Izdatel'stvo, 1956); G. V. Platonov, *Darvin, darvinizm i filosofiia* (Moscow: Izdatel'stvo Politicheskoi Literatury, 1959), pp. 192, 198, 202; N. V. Lebedev, *Lektsii po darvinizmu* (Moscow: Moscow University, 1962), pp. 284–88); V. A. Alekseev, *Osnovy darvinizma* (Moscow: Moscow University, 1964), p. 348. The anti-Malthusianism of Russian biologists was also emphasized in Lysenkoist biographies, which are cited in the separate chapters of the present volume. For an example of the struggle between Lysenkoists and their opponents over the legacy of prerevolutionary Russian evolutionism, see *The Situation in Biological Science* (New York: International Publishers, 1949). For recent historical discussions by Soviet authors, see especially Ia. M. Gall's rich study *Bor'ba za sushchestvovanie kak faktor evoliutsii* (Leningrad: Nauka, 1973) and K. M. Zavadskii, *Razvitie evoliutsionnoi teorii posle Darvina* (Leningrad: Nauka, 1973), pp. 152–64. Relevant Western works are the essays by James Allen Rogers listed in Chapter 2, note 10; Alexander Vucinich, *Science in Russian Culture, 1861–1917* (Stanford, Calif.: Stanford University Press, 1970), p. 274; and Francesco M. Scudo and Michele Acanfora, "Darwin and Russian Evolutionary Biology," in David Kohn, ed., *The Darwinian Heritage* (Princeton, N.J.: Princeton University Press, 1985), pp. 731–54.

3. Among the important naturalists that I do not address, at least three fit the pattern: botanist A. S. Famintsyn and zoologists N. A. Kholodkovskii and L. S. Berg. See A. L. Kursanov and S. R. Mikulinskii, eds., *Andrei Sergeevich Famintsyn* (Leningrad: Nauka, 1981); A. S. Famintsyn, "Darvin i Ego znachenie v biologii," *Otechestvennye zapiski* 3 (1874): 129–50; A. S. Famintsyn, "O roli simbioza v evoliutsii organizmov," *Izvestiia Akademii Nauk* 6, No. 1 (1912): 51–68, and 6, No. 11 (1912): 707–14; N. A. Kholodkovskii, *Biologicheskie ocherki* (Moscow-Petrograd, 1923); Leo S. Berg, *Nomogenesis or Evolution Determined by Law* [1922] (Cambridge, Mass.: MIT Press, 1969).

Chapter 1

1. Charles Darwin, *On the Origin of Species* (New York: D. Appleton and Company, 1876), p. 63. This phrase first appeared in the third edition, published in 1861.

2. Charles Darwin, *On the Origin of Species*, a facsimile of the first edition (Cambridge, Mass.: Harvard University Press, 1964), p. 62.

3. Ibid., pp. 62–63.

4. Ibid., p. 63. The last sentence here reads in full, "It is the doctrine of Malthus applied with manifold force to the whole animal and vegetable kingdoms; for in this case there can be no artificial increase of food, and no prudential restraint from marriage." In his introduction Darwin noted simply, "This is the doctrine of Malthus, applied to the whole animal and vegetable kingdoms" (p. 5).

5. Ibid., p. 78.

6. Ibid., p. 127.

7. He also invoked the use and disuse of parts, the inheritance of acquired characteristics, and isolation. In subsequent editions of the *Origin*, these factors acquired a greater emphasis, but natural selection remained the primary factor in evolution.

8. Darwin, *Origin of Species*, p. 69.

9. Ibid., p. 67.

10. Ibid., pp. 68–69.

11. Ibid., p. 205.

12. The final paragraph of the *Origin* begins, "It is interesting to contemplate an entangled bank, clothed with many plants of many kinds, with birds singing on the bushes, with various insects flitting about, and with worms crawling through the damp earth, and to reflect that these elaborately constructed forms, so different from each other, and dependent on each other in so complex a manner, have all been produced by laws acting around us" (p. 489).

13. Charles Darwin, *A Naturalist's Voyage: Journal of Researches into the Natural History and Geology of the countries visited during the voyage of H.M.S. 'Beagle' round the world*, in Paul H. Barrett and R. B. Freeman, eds., *The Works of Charles Darwin* (New York: New York University Press, 1986), vol. 3, pp. 462–63.

14. This is discussed in the conclusion to the present volume.

15. Darwin, *Origin of Species*, p. 75.
16. Ibid., p. 76.
17. Ibid., p. 79.
18. It is incessant in the sense that natural selection operates at all times in the organism's existence, although, as Darwin pointed out, the struggle for existence was more intense in certain periods and at certain times in an organism's life. See Darwin, *Origin of Species*, pp. 66–67.
19. Ibid., p. 236.
20. Charles Robert Darwin, *The Descent of Man, and Selection in Relation to Sex*, a facsimile of the first English edition, (Princeton, N.J.: Princeton University Press, 1981), pp. 80–81.
21. Ibid., p. 82.
22. Darwin, *Origin of Species*, p. 488.
23. Darwin, *Descent of Man*, p. 185.
24. Howard Gruber interprets Darwin's self-description succinctly:

The appellation *liberal-radical* probably meant then approximately what it would mean today—concerned more with the preservation and extension of individual liberty than with the preservation of hallowed social institutions; concerned more with human rights than with property rights; favorably yet cautiously disposed toward social change; unattached to any organized group that would pursue the desired aims in a manner disturbing to the comfort and tranquility of upper-middle-class life.

Howard E. Gruber, *Darwin on Man: A Psychological Study of Scientific Creativity*, 2d ed. (Chicago: University of Chicago Press, 1981), p. 69.

25. Darwin, *Descent of Man*, p. 180.
26. Ibid., pp. 403–4.
27. Patricia James, *Population Malthus: His Life and Times* (London, Boston, and Henley: Routledge & Kegan Paul, 1979).
28. Thomas Robert Malthus, *An Essay on the Principle of Population* [1798], in E. A. Wrigley and David Souden, eds., *The Works of Thomas Robert Malthus* (London: William Pickering, 1986), vol. 1, p. 9. Among the "preventive checks" were "vicious customs with regard to women"—abortion and infanticide.
29. Ibid., p. 14.
30. Ibid., p. 21.
31. Ibid., p. 22.
32. "It cannot be true, therefore," Malthus added, "that among animals, some of the offspring will possess the desirable qualities of the parents in a greater degree; or that animals are indefinitely perfectible" (Ibid., pp. 60–61).
33. Ibid., p. 8.
34. Ibid., p. 10.
35. See E. A. Wrigley and David Souden, eds., *The Works of Thomas Robert Malthus* (London: William Pickering, 1986), vols. 2 and 3. These very useful volumes offer the sixth edition of Malthus's *Essay* with variant readings from the second edition.
36. M. Blaug, "Introduction," in Thomas R. Malthus, *An Essay on the Principle of Population* [1803] (Homewood, Ill.: Richard D. Irwin, Inc., 1963), p. viii.
37. Robert Young, "Malthus and the Evolutionists: The Common Context of Biological and Social Theory," *Past and Present* 43 (1969): 114.
38. Blaug, "Introduction," pp. viii–ix. For differing perspectives on Malthus and his reception, see John Cunningham Wood, ed., *Thomas Robert Malthus: Critical Assessments*, 4 vols. (London: Croon Helm, 1986).
39. Edward Manier, *The Young Darwin and His Cultural Circle: A Study of Influences Which Helped Shape the Language and Logic of the First Drafts of the Theory of Natural Selection* (Dordrecht and Boston: D. Reidel Publishing Company, 1978), p. 20.
40. R. C. Stauffer, ed., *Charles Darwin's Natural Selection* (Cambridge: Cambridge University Press, 1975), p. 176.
41. Ibid., p. 176.
42. Francis Darwin, ed., *Life and Letters of Charles Darwin* (New York: D. Appleton and Company, 1891), vol. 2, p. 111.

43. Ibid., p. 230.

44. Charles Darwin, *Autobiography*, [1876], ed. Nora Barlow, (London: Collins, 1958), p. 120.

45. Alfred Russel Wallace, *The Wonderful Century* (London: Swan Sonnenschein & Company, 1898), p. 139. See also H. Lewis McKinney, *Wallace and Natural Selection* (New Haven, Conn., and London: Yale University Press, 1972). For Wallace's subsequent second thoughts about Malthus and their effect on his evolutionary ideas, see Young, "Malthus and the Evolutionists," pp. 132–34.

46. T. H. Huxley, *On the Origin of Species; or, The Causes of the Phenomena of Organic Nature* (New York: D. Appleton and Company, 1873), pp. 120, 123–24.

47. Leonard Huxley, ed., *Life and Letters of Sir Joseph Hooker*, (London: Murray, 1918), vol. 2, p. 43. Hooker easily applied the selection theory to the relative virtues of Britain's social classes:

> I have no hesitation in thinking that the honor of our uppermost Tory classes of a higher order than of the middle, just as their vices are more conspicuous. They can *afford* to be more high-minded, just as they can *afford* to commit sins that damn a lower class, and upon the whole I expect they are less vicious than the middle class and infinitely less than the lower. Indeed upon scientific grounds I have stated before (Natural selection, and continued success only attending honesty) I think it should be so.

His correspondent in the United States, Asa Gray, was in complete accord:

> Curiously enough, Mr. Darwin's theory is grounded upon the doctrine of Malthus and the doctrine of Hobbes. The elder DeCandolle had conceived the idea of the struggle for existence, and in a passage which would have delighted the cynical philosopher of Malmesbury, had declared that all Nature is at war, one organism with another or with external Nature; and Lyell and Herbert had made considerable use of it. But Hobbes in his theory of society, and Darwin in his theory of natural history, alone have built their systems upon it.

Gray had no objection to Darwin's "thousandfold confirmation and extension of the Malthusian doctrine," and confessed that "in fact, we began to contract a liking for a system which at the outset illustrates the advantages of good breeding, and which makes the most of every creature's best." Asa Gray, *Darwiniana: Essays and Reviews Pertaining to Darwinism*, ed. A. Hunter Dupree, (Cambridge, Mass.: Harvard University Press, 1963), pp. 30, 34.

48. The following discussion has benefitted especially from an incisive review of this literature by Antonello La Vergata, "Images of Darwin: A Historiographic Overview," in David Kohn, ed., *The Darwinian Heritage* (Princeton, N.J.: Princeton University Press, 1985), pp. 953–57.

49. J. D. Bernal, *Science in History* (Cambridge, Mass.: MIT Press, 1971), vol. 4, p. 1233.

50. Young, "Malthus and the Evolutionists," p. 111.

51. Dov Ospovat, *The Development of Darwin's Theory: Natural History, Natural Theology, and Natural Selection, 1838–1859* (Cambridge: Cambridge University Press, 1981), p. 233.

52. Sylvan S. Schweber, "The Wider British Context in Darwin's Theorizing," in David Kohn, ed., *The Darwinian Heritage* (Princeton, N.J.: Princeton University Press, 1985), p. 38.

53. Ernst Mayr, *The Growth of Biological Thought: Diversity, Evolution, and Inheritance* (Cambridge, Mass., and London: Harvard University Press, 1982), pp. 492–93.

54. Gruber, *Darwin on Man*, p. 174.

55. Sandra Herbert, "Darwin, Malthus, and Selection," *Journal of the History of Biology* 4, 1 (1971): 209–17.

56. David Kohn, "Theories to Work By: Rejected Theories, Reproduction, and Darwin's Path to Natural Selection," *Studies in History of Biology* 4 (1980): 144.

57. Kohn, "Theories to Work By," 145–46.

58. M. J. S. Hodge and David Kohn, "The Immediate Origins of Natural Selection," in Kohn, *The Darwinian Heritage*, p. 195. See also M. J. S. Hodge, "The Development of Darwin's General Biological Theorizing," in D. S. Bendall, ed., *Evolution from Molecules to Men* (Cambridge: Cambridge University Press, 1983), pp. 43–62.

59. Young, "Malthus and the Evolutionists," p. 129.

60. See especially Robert M. Young, "The Historiographic and Ideological Contexts of the Nineteenth Century Debate on Man's Place in Nature," in his *Darwin's Metaphor: Nature's Place in Victorian Culture* (Cambridge: Cambridge University Press, 1985), pp. 164–247, and "Darwinism Is Social," in Kohn, *The Darwinian Heritage*, pp. 609–38.

61. For a discussion of the cultural sources, interaction, and significance of Darwin's metaphors, see Manier, *The Young Darwin and His Cultural Circle*; and Howard Gruber, "The Evolving Systems Approach to Creative Scientific Work: Charles Darwin's Early Thought," in Thomas Nickles, ed., *Scientific Discovery: Case Studies* (Dordrecht, Boston, and London: D. Reidel Publishing Company, 1980), pp. 113–30. The term "figures of thought" is Gruber's.

62. Gruber, "The Evolving Systems Approach," p. 127.

63. Manier, *The Young Darwin and His Circle*, p. 82.

64. Ibid., p. 192. See also Peter Bowler, "Malthus, Darwin, and the Concept of Struggle," *Journal of the History of Ideas* 37 (1976): 631–50.

65. Nicholas V. Riasanovsky, *A History of Russia*, 2d ed. (New York: Oxford University Press, 1969), p. 5.

66. P. Belov, *Uchebnik geografii Rossiiskoi Imperii* (St. Petersburg, 1875), pp. 3, 78. According to one modern historian of population, European Russians numbered 40 million in 1800, 57 million in 1850, and 100 million in 1900. This compared with Great Britain's population of 11 million in 1800, 21 million in 1850, and 37 million in 1900. European Russia's annual population growth, then, was 0.71 percent from 1800 to 1850, rising to 1.14 percent from 1850 to 1900. Great Britain's rate was 1.30 percent (almost twice that of European Russia) from 1800 to 1850 and 1.14 percent (the same as European Russia) from 1850 to 1900. See E. A. Wrigley, *Population and History* (New York and Toronto: McGraw-Hill, 1969).

67. Unlike France, England, and Prussia, Russia was not plagued by a landless peasantry, and Russian commentators did not attribute the problem of "land hunger" to an inadequate amount of arable land. This was clear in the solutions proposed most frequently by state officials: expanding the amount of land that peasants were permitted to purchase and rent (by making available some of the vast holdings of the government and nobility) and encouraging peasants to resettle to the east. As for the peasants themselves, rather than abandoning the age-old three-field system, which left one-third of available land fallow each season, they pressed for *prostor*, for greater access to the great open spaces of Mother Russia. See Geroid Tanquary Robinson, *Rural Russia under the Old Regime* (Berkeley: University of California Press, 1969), especially pp. 94–101. For the perceptions of tsarist officials, see David A. J. Macey, *Government and Peasant in Russia, 1861–1906* (DeKalb, Ill.: Northern Illinois University Press, 1987), especially pp. 18–19, 58–59, 79–80, 122–23, 133–34, 226–31.

68. Jerome Blum, *Lord and Peasant in Russia* (Princeton, N.J.: Princeton University Press, 1961); Alfred Rieber, *Merchants and Entrepreneurs in Imperial Russia* (Chapel Hill, N.C.: University of North Carolina Press, 1982); Reginald Zelnick, *Labor and Society in Tsarist Russia* (Stanford, Calif.: Stanford University Press, 1971).

69. August von Haxthausen, *Studies on the Interior of Russia*, edited with an introduction by S. Frederick Starr; trans. Eleanore L. M. Schmidt (Chicago and London: University of Chicago Press, 1972), p. xxx.

70. Haxthausen, *Studies*, p. xxxi. See also Georg Brandes, *Impressions of Russia* [1889], with an introduction by Richard Pipes (New York: Crowell Co., 1966). Echoing Russian populists, Brandes portrayed Russian communality as a response to external dangers:

> Is it not evident that in such circumstances there is no use for the saying, "My house is my castle"? Such an idea would be madness. Here, it is not only unsuited to the surroundings, but impossible, to live alone, each family and each farm by itself. Each one daily needs the help of his neighbors for protection against the Tatars, for defence against the wild beasts, for the clearing of the woodland, and for breaking up the soil (pp. 23–24).

71. See Nancy Mandelker Friedan, *Russian Physicians in an Era of Reform and Revolution* (Princeton, N.J.: Princeton University Press, 1981); and Daniel P. Todes, "Biological Psychology and the Tsarist Censor: The Dilemma of Scientific Development," *Bulletin of the History of Medicine* 58 (1984): 529–44.

72. Cited and translated in Alexander Vucinich, "Russia: Biological Sciences," in Thomas F. Glick, ed., *The Comparative Reception of Darwinism* (Austin, Tex., and London: University of Texas Press, 1972), pp. 229–30.

73. M. A. Antonovich, *Charl'z Darvin i ego teoriia* (St. Petersburg, 1896), pp. 233–34. In the years 1859–1890 theological journals did not print a single systematic critique of Darwin's theory,

although some theologians contributed critical reviews to other publications after the appearance of *Descent of Man*. See George L. Kline, "Darwinism and the Russian Orthodox Church," in Ernest J. Simmons, ed., *Continuity and Change in Russian and Soviet Thought* (Cambridge, Mass.: Harvard University Press, 1955), pp. 307–28.

74. See, for example, K. M. Zavadskii, *Razvitie evolutionnoiteorii posle Darvina* (Leningrad: Nauka, 1973), pp. 137–74; and Vucinich, "Russia: Biological Sciences," pp. 227–55.

Chapter 2

1. K. A. Timiriazev, "Zemledelie i fiziologiia rasteniia" [1906], *Sochinenie* (Moscow, 1937), vol. 3, p. 31.
2. P. N. Tkachev, "Nauka v poezii i poeziia v nauke" [1870], *Sochinenie v dvukh tomakh* (Moscow, 1975), vol. 1, p. 398.
3. Thomas Robert Malthus, *An Essay on the Principle of Population* [1803, 1826], in E. A. Wrigley and David Souden, eds., *The Works of Thomas Robert Malthus* (London: David Pickering, 1986), vol. 2, pp. 103–4. As is indicated in this concordance edition, the citations that follow appear sometimes in the second edition, sometimes in the sixth, and sometimes in both.
4. Ibid., p. 103.
5. Ibid., pp. 106–8.
6. Ibid., p. 188.
7. Malthus's *Essay* was translated into German in 1807 and into French in 1809. Extensive excerpts appeared in a Spanish journal in 1808, and the complete text was translated in 1845. The first Russian translation was T. R. Mal'tus', *Opyt o zakone narodonaseleniia*, trans. P. A. Bibikov (St. Petersburg, 1868).
8. T. R. Mal'thus, *Opyt' zakona o narodonaselenii*, trans. I. A. Vernera (Moscow, 1895). For Russian criticisms of Malthus, see G. N. Korostelev, *Kritika mal'tuzianskikh i neomal'tuzianskikh vzgliadov: Rossiia XIX nachala XX v.* (Moscow: Statistika, 1978). For an interesting commentary on Malthus and the weaknesses of existing Russian translations of his *Essay*, see M. V. Ptukha, "Uchenie o narodonaselenii T. R. Mal'tusa," *Demograficheskie Tetrady* 4–5 (1972): 138–91.
9. That Russians were interested in Malthus primarily as a political figure is apparent from their silence regarding his analysis of population and food supply in their own country. Nor, so far as I can tell, did a single Russian consider the changes in the second edition of his *Essay* to be more than a rhetorical device. Even P. A. Bibikov's essay on Malthus, which introduced his translation of the fifth edition of the *Essay*, dealt with the "pure form" of Malthus's argument presented in 1798.
10. James Allen Rogers deals with the reception of Darwinism by Russian social thinkers in a series of articles, several of which mention the anti-Malthusian themes in their comments. See his "Darwinism, Scientism and Nihilism," *The Russian Review* 19, No. 1 (1960): 10–23; "The Russian Populists' Response to Darwin," *Slavic Review* 22, No. 3 (1963): 456–68; "Marxist and Russian Darwinism," *Jahrbücher für Geschichte Osteuropas* 13, No. 2 (1965): 199–211; "Proudhon and the Transformation of Russian Nihilism," *Cahiers du monde Russe et Soviétique* 13, No. 4 (1972): 514–23; "Darwinism and Social Darwinism," *Journal of the History of Ideas* 33 (1972): 265–80.
11. *Syn otechestva* 17 (1818): 195.
12. K. F. German, *Statisticheskie issledovaniia otnositel'no Rossiiskoi Imperii: O narodonaselenii* (St. Petersburg, 1819), Part 1, p. 48.
13. K. I. Arsen'ev, "Izsledovaniia o chislennom otnoshenii polov' v narodonaselenii Rossii," *Zhurnal Ministerstva Vnutrennykh Del* 5, No. 1 (1844): 9.
14. Ibid.
15. Ibid., 8.
16. Ibid., 9.
17. A. I. Butovskii, *Opyt o narodnom bogatstve, ili o nachalakh politicheskoi ekonomii* (St. Petersburg, 1847), p. 350.
18. Ibid., pp. 367, 370.

19. Ibid., p. 376.
20. A. I. Butovskii, "Obshchinnoe vladenie i sobstvennost'," *Russkii vestnik* 16 (1858): 34.
21. V. A. Miliutin, "Mal'tus' i ego protivniki," *Sovremennik* 4, No. 8 (1847): 135.
22. V. A. Miliutin, "Mal'tus i ego protivniki," *Sovremennik* 4, No. 9 (1847): 49–50.
23. Miliutin, "Mal'tus," No. 8, p. 151.
24. Miliutin, "Mal'tus," No. 9, pp. 42, 54–55, 61–62.
25. Miliutin, "Mal'tus," No. 8, pp. 169, 172.
26. Ibid., 169.
27. Ibid., 171.
28. V. F. Odoevsky, *Russian Nights* [1844], trans. Olga Koshansky-Olienikov and Ralph Mitlow (New York: E. P. Dutton and Company, 1965) p. 51.
29. Ibid., pp. 50–51.
30. Ibid., p. 70.
31. Ibid., p. 91.
32. Ibid., p. 51.
33. Commenting on the 1860s, Reginald Zelnick has pointed out that "Malthusian pessimism was not a significant phenomenon in Russia." See *Labor and Society in Tsarist Russia* (Stanford, Calif.: Stanford University Press, 1971) p. 114.
34. For a liberal defense of Malthus, see Iu. Zhukovskii, "Vopros narodonaseleniia," *Vestnik Evropy* 1 (1871): 167–205.
35. For a discussion of Chernyshevskii's critique of Malthus, see Korostelev, *Kritika mal'tuzianskikh*, pp. 67–82.
36. N. G. Chernyshevskii, *Polnoe sobranie sochineniia* (Moscow: Khudozhestvennaia Literatura, 1950), vol. 9, p. 309.
37. Dmitry Pisarev, *Selected Philosophical, Social and Political Essays* (Moscow: Foreign Languages Publishing House, 1958), p. 43.
38. Ibid., pp. 202–3.
39. Ibid., p. 208.
40. *Russkoe slovo* 7 (1865): 5.
41. *Russkoe slovo* 1 (1865): 84.
42. A. I. Gertsen, *Sobranie sochinenie v tridtsati tomakh* (Moscow, 1955), vol. 6, p. 163.
43. Tkachev, "Nauka v poezii," p. 398. See Deborah Hardy's excellent biography, *Petr Tkachev* (Seattle and London: University of Washington Press, 1977).
44. For a similar problem with physiological psychology, see Daniel P. Todes, *From Radicalism to Scientific Convention: Biological Psychology in Russia from Sechenov to Pavlov*, doctoral dissertation, University of Pennsylvania, 1981.
45. Cited in G. V. Plekhanov, "N. G. Chernyshevskii," *Izbrannye filosofskie proizvedeniia* (Moscow: Sotsial'no-ekonomicheskaia Literatura, 1958), vol. 4, p. 79.
46. "Bibliograficheskii listok," *Russkoe slovo* 5 (1863): 72.
47. See Daniel P. Todes, "Biological Psychology and the Tsarist Censor: The Dilemma of Scientific Development," *Bulletin of the History of Medicine* 58 (1984): 529–44.
48. *Russkoe slovo* 3 (1865): 254.
49. Ibid.
50. Pisarev, *Selected Essays*, p. 305.
51. Ibid., p. 332.
52. Ibid., pp. 330–31.
53. *Russkoe slovo* 8 (1864): 94.
54. See Rogers, "The Russian Populists' Response to Darwin," 457–58; M. A. Antonovich, "Lzherealisty," *Sovremennik* 7 (1865): 54; and Pisarev, *Sochinenie* (St. Petersburg, 1897), vol. 5, pp. 143–44. Antonovich favorably reviewed Darwin's theory, including his use of Malthus, in "Teoriia proiskhozhdeniia vidov v tsarstve zhivotnom," *Sovremennik* 10, No. 3 (1864): especially pp. 76–77. *The Contemporary* was abandoning scientism and adopting attitudes that foreshadowed the populism of subsequent decades.
55. In a review of P. A. Bibikov's *Kriticheskie etiudi*, Zaitsev insisted that he had not endorsed slavery. He had simply contended that humans must master and use the laws uncovered by Dar-

win if they were to eliminate war, inequality, and exploitation. This article of 1865 is reprinted in V. A. Zaitsev, *Izbrannye sochineniia* (Moscow: Vsesoiuz. Ob. Politkatorzhan i Ssylno-poselentsev, 1934), pp. 428–42.

56. See A. E. Gaisinovich, "Biolog-shestidesiatnik N. D. Nozhin i ego rol' v razvitii embriologii i darvinizma v Rossii," *Zhurnal obshchei biologii* 13, No. 5 (1952): 377–87.

57. N. D. Nozhin, "Nasha nauka i uchenye: Stat'ia V," *Knizhnyi vestnik* 6 (1866): 174.

58. Ibid., 176.

59. Ibid.

60. Ibid.

61. Ibid.

62. Ibid., 177.

63. N. D. Nozhin, "Po povodu statei 'Russkogo Slova' o nevol'nichestve," *Iskra* 8 (1865): 115.

64. Rogers concludes that Zaitsev "was the only thinker of any note who raised his voice in favor of what later could be construed as a European type of Social Darwinism"; see "Darwinism, Scientism and Nihilism," 21. R. A. Bashmakova agrees that "unlike the West, there did not develop in Russia a general social-Darwinist school," but finds elements of "Social Darwinism" in the thought of T. D. Faddeev, S. A. Muromtsev, Iu. G. Zhukovskii, P. F. Lilienfel'd, and others; see "Sotsial'nyi darvinizm v Rossii i ego kritika," *Problemy sotsiologii narodonaseleniia* 234, No. 2 (1974): 3–22.

65. P. A. Bibikov, *Kriticheskie etiudi* (St. Petersburg, 1865). For biographical information see "Bibikov, Petr Alekseevich," *Russkii biograficheskii slovar'* [St. Petersburg, 1908], vol. 3, reprinted by Kraus Reprint Corporation, New York, 1962, p. 31.

66. T. R. Mal'tus, *Opyt o zakone narodonaseleniia*, trans. P. A. Bibikov, t. I, *s premechaniami i so statei perevodchika, zhizn' i trudy Mal'tusa* (St. Petersburg, 1868), pp. 6, 85.

67. Bibikov, "Zhizn' i trudy Mal'tusa," in Mal'tus, *Opyt o zakone narodonaseleniia*, p. 85.

68. The word *stremliat'sia* can be translated as either "to tend" or "to aspire." Here Bibikov clearly is using it in the former sense, following Malthus's argument about the "tendency" of organisms to multiply their numbers in a geometrical progression.

69. Bibikov, "Zhizn' i Trudy Mal'tusa," p. 22.

70. Ibid., pp. 21–22.

71. Ibid., p. 23.

72. Ibid., p. 24.

73. Ibid., pp. 24–25. Compare this passage with the citation from Darwin's *Origin of Species* in Chapter 1, p. 8.

74. Bibikov, "Zhizn' i trudy Mal'tusa," p. 24.

75. Ibid., p. 25.

76. Ibid.

77. Ibid., p. 26.

78. Ibid., pp. 31–32.

79. Ibid., pp. 28–29.

80. Ibid., p. 34.

81. Ibid.

82. Ibid., p. 35.

83. Ibid., p. 44.

84. See the review of Bibikov's essay in the liberal *Vestnik Evropy* 5 (1869): 907.

85. S. A. Vengerov, *Ocherki po istorii russkoi literatury*, 2d ed. (St. Petersburg, 1907), p. 85.

86. N. V. Shelgunov, "Progressivnaia reaktsiia," *Delo* 4 (1879): 148.

87. N. V. Shelgunov, "Obrazovannyi proletariat" [1869], *Sochinenie* (St. Petersburg, 1862), p. 165.

88. For a review of populist comments about the rise of capitalism in the countryside, see N. P. Emel'ianov, *Otechestvennye zapiski N. A. Nekrasova (1868–1877)* (Leningrad: Leningrad University Press, 1977). The actual degree of capitalist development at this time is a matter of debate among historians, but the populist preoccupation with this threat is undeniable.

89. N. K. Mikhailovskii, "Zapiski profana: O demokratizme estestvennykh nauk" [1875], *Sochinenie* (St. Petersburg, 1897), vol. 3, p. 284.

90. Ibid.
91. Ibid., p. 286.
92. N. K. Mikhailovskii, "Analogicheskii metod v obshchestvennoi nauke," *Otechestvennye zapiski* 7 (1869): 44.
93. Ibid.
94. Ibid., 44-45.
95. N. K. Mikhailovskii, "Teoriia Darvina i obshchestvennaia nauka," *Sochinenie* (St. Petersburg, 1896), vol. 1, p. 294.
96. Mikhailovskii, "Teoriia Darvina," p. 295. Mikhailovskii did not distinguish between "Darwinism" and "Social Darwinism"; he used the single term "Darwinism"for the view that the struggle for existence between individuals was the source of progress.
97. P. L. Lavrov, "Sotsializm i bor'ba za sushchestvovanie" [1873], in I. A. Teordorovich, ed., *Izbrannye sochineniia* (Moscow, 1935), vol. 4, pp. 99-110. Lavrov had earlier written extensively about the relationship of the natural and social sciences. For his criticism of Pisarev's scientism, see A. Ugriumov, "O publitsistakh-populiarizatorakh i estestvoznanie," *Sovremennik* 9 (1865). (There is some question as to the authorship of this article, but internal evidence convinces me that it was indeed Lavrov.) See also his *Historical Letters* [1868-1869], trans and ed. James Scanlan (Berkeley, Calif.: University of California Press, 1967).
98. Lavrov, "Sotsializm i bor'ba," p. 101.
99. Ibid.
100. Ibid.
101. Ibid., 102.
102. Ibid., 103.
103. Ibid.
104. Ibid., 110. Note that Lavrov distinguished between Darwin, with his great scientific contribution, and "Darwinists," who distorted the contribution's significance.
105. See Engels's letter of November 12-17, 1875, in *Perepiska K. Marksa i F. Engel'sa s russkimi politicheskimi deiateliami* (Moscow, 1947), pp. 170-72.
106. *Chernyshevskii v Sibiri: Perepiska s rodnymi* (St. Petersburg, 1913), vol. 1, p. 69.
107. Ibid., vol. 3, p. 18.
108. Ibid., vol. 1, p. 71.
109. Ibid., vol. 3, p. 18.
110. Ibid.
111. Ibid.
112. For Timiriazev's reaction see Chapter 8, pp. 162-63.
113. Tkachev, *Sochinenie*, vol. 1, pp. 279, 285-86.
114. Tkachev, "Nauka v poezii," p. 398.
115. Ibid.
116. Ibid., pp. 398-99.
117. Ibid., p. 339.
118. Ibid., p. 400.
119. Ibid., p. 399.
120. L. I. Mechnikov, "Kolonizatsiia v Avstralii i v Amerike," *Delo* 12 (1880); "Proekty agrarnykh preobrazovanii v Anglii," *Delo* 1 (1883).
121. L. I. Mechnikov, "Shkola bor'by v sotsiologii," *Delo* 1 (1884): 28-64.
122. Ibid., 29.
123. Ibid., 39.
124. Ibid., 36.
125. Ibid., 53.
126. G. V. Plekhanov, *Izbrannye filosofskie proizvedeniia* (Moscow: Sotsial'no-ekonomicheskaia Literatura, 1956), vol. 1, pp. 690-91. This was also Marx and Engels's position. In 1878 Engels wrote:

> However great the blunder made by Darwin in accepting the Malthusian theory so naively and uncritically, nevertheless anyone can see at the first glance that no Malthusian spectacles are required to perceive the struggle for existence in nature—the contradiction between the countless

host of germs which nature so lavishly produces and the small number of those which ever reach maturity, a contradiction which in fact for the most part finds its solution in a struggle for existence—often of extreme cruelty.

Darwin's theory had also stimulated studies of the laws of population, "which have been left practically uninvestigated." Frederick Engels, *Anti-Duhring* (Moscow: Progress Publishers, 1969), p. 86.

127. G. V. Plekhanov, *Izbrannye filosofskie proizvedeniia* (Moscow: Sotsial'no-ekonomicheskaia Literatura, 1958), vol. 4, pp. 279-87.

128. Ibid., pp. 283-84.

129. Ibid., pp. 285-86.

130. S. A. Rachinskii, "Tsvety i nasekomye," *Russkii vestnik* 1 (1863).

131. See L. Ia. Kharakhorin, "Charl'z Darvin i tsarskaia tsenzura," *Trudy Instituta Istorii Estestvoznaniia i Tekhniki* 30 (1962): 6; and M. I. Vladislavlev, "Sovremennyi materializm," *Epokha* 1 (1865).

132. See A. Gusev, *Naturalist Uolles, ego russkie perevodchiki i kritiki* (Moscow, 1879); and I. F. Tsion, *Nigilisty i nigilizm* (Moscow, 1886). For the account of a defender of Darwin, see M. A. Antonovich, *Charl'z Darvin i ego teoriia* (St. Petersburg, 1896), pp. 232-51.

133. See Linda Gerstein, *Nikolai Strakhov* (Cambridge, Mass.: Harvard University Press,1971). Strakhov's articles on natural science were compiled in *O metode estestvennykh nauk i znachenii ikh v obshchem obrazovanii* (St. Petersburg, 1865).

134. N. N. Strakhov, "Durnye znaki," *Vremia* 11 (1862): 167.

135. Ibid., 168-69.

136. Ibid., 169.

137. Gerstein, *Strakhov*, p. 156. Like most Russian criticisms of a social Darwinist, Strakhov's was directed at a foreign author.

138. N. N. Strakhov, *Bor'ba s Zapadom v nashei literature* (St. Petersburg, 1882), p. 121.

139. See Robert E. MacMaster, *Danilevsky: A Russian Totalitarian Philosopher* (Cambridge, Mass.: Harvard University Press, 1967); Iu. Ivask, "Filosofiia istorii Danilevskogo v knige *Rossii i Evropa*," in N. Ia. Danilevskii, ed., *Rossiia i Evropa* [1869], (New York and London: Johnson Reprint Corporation, 1966), p. v; and N. Ia. Danilevskii, *Sbornik politicheskikh i ekonomicheskikh statei* (St. Petersburg, 1890).

140. Danilevskii, *Rossiia i Evropa*, p. 146.

141. Ibid., pp. 146-47.

142. Ibid., p. 147.

143. N. Ia. Danilevskii, *Darvinizm: Kriticheskoe issledovanie* (St. Petersburg: Komarov, 1885), vol. 1, p. 478.

144. Ibid.

145. Ibid., p. 461.

146. Ibid., pp. 452-53.

147. Ibid., p. 461.

148. Ibid., pp. 483-84.

149. Ibid., p. 479.

150. N. N. Strakhov, "Polnoe oproverzhenie darvinizma," *Russkii vestnik* 1 (1887): 9-62; K. A. Timiriazev, "Oprovergnut-li darvinizm?" *Russkaia mysl'* 5-6 (1887): 145-80; N. N. Strakhov, "Vsegdashnaia oshibka darvinistov," *Russkii vestnik* 11-12 (1887): 98-129; A. S. Famintsyn, "N. Ia. Danilevskii i darvinizm," *Vestnik Evropy* 2 (1889): 616-43; N. N. Strakhov, "A. S. Famintsyn o 'darvinizme' N. Ia. Danilevskogo," *Russkii vestnik* 4 (1889): 225-43; K. A. Timiriazev, "Strannyi obrazchik nauchnoi kritiki," *Russkaia mysl'* 3 (1889): 90-102; and K. A. Timiriazev, "Bezsilnaia zloba antidarvinista," *Russkaia mysl'* 5 (1889): 17-52; 6 (1889): 65-82; 7 (1889): 58-78.

151. Cited in Gerstein, *Strakhov*, p. 162.

152. K. P. Pobedonostsev, *Reflections of a Russian Statesman*, trans. R. C. Long (Ann Arbor, Mich.: University of Michigan Press, 1965), p. 183.

153. N. Rumiantsev, "Darvinizm (kriticheskoe issledovanie)," *Vera i razum* 2 (1895): 1-31, 135-70, 291-302, 488-504; 2 (1896): 25-45, 90-110, 139-60, 240-68.

154. M. Glubokovskii, "K voprosu o darvinizme," *Vera i razum* 2 (1892): 60-65. See George L. Kline, "Darwinism and the Russian Orthodox Church," in Ernest J. Simmons, ed., *Continuity*

and Change in Russian and Soviet Thought, (Cambridge, Mass.: Harvard University Press, 1955), pp. 307–28.

155. Lev Tolstoi, *What to Do?*, trans. Isabel Hapgood (New York: Crowell and Company, 1887), p. 172.

156. Ibid., p. 173.

157. Ibid.

158. Ibid., p. 175.

159. Ibid., p. 176.

160. L. D. Opul'skaia. *Lev Nikolaevich Tolstoi: Materialy k biografii s 1886 do 1892 god.* (Moscow: Nauka, 1979), p. 146.

161. For Timiriazev's response see Chapter 8, p. 162.

162. *Tolstoy's Letters*, trans. and ed. R. F. Christian. (New York: Charles Scribner's Sons, 1978), vol. 2, 1880–1910, p. 717.

Chapter 3

1. A. N. Beketov, "Nravstvennost' i estestvoznanie," *Voprosy filosofii i psikhologii* 6, No. 2 (1891): 60.

2. This citation and translation are adapted from Konstantin Mochulsky, *Aleksandr Blok*, trans. Doris V. Johnson (Detroit: Wayne State University Press, 1983), p. 19.

3. For biographies of Beketov see A. A. Shcherbakova, *Andrei Nikolaevich Beketov* (Moscow, 1958); and B. E. Raikov, *Russkie biologi-evoliutsionisty do Darvina* (Moscow and Leningrad, 1959). See also Beketov's autobiographical sketch, "Beketov, Andrei Nikolaevich, botanik," in S. A. Vengerov, ed., *Kritiko-biograficheskii slovar' russkikh pisatelei i uchenykh* (St. Petersburg, 1891), vol. 2, vyp. 22–30, pp. 353–63; and the commemorative volume, *Trudy Sankt-Peterburgskogo Obshchestva Estestvoispytatelei* 33 (1903), 1.

4. For a description of the Petrashevskii circle, see Franco Venturi, *The Roots of Revolution* (New York: Grosset and Dunlap, 1960), pp. 79–89. One historian describes Fourierism as the protest of "the petty-bourgeois consumer, the small employee or artisan, both against the absence of means for insuring his security in the existing undemocratic order and against a nascent capitalism." Fourier's phalanstery offered a means of abolishing the distinction between consumer and producer "and organizing society into manageable, self-sufficient units (like the old village), where each man will work only at what he pleases, in 'harmony' and security, not competition, with his fellows. Thus Fourierism is a democratic glance backward to an idealized agrarian order as much as a glance forward to the problems of modern industry; in this it bears the unmistakable imprint of conditions that are still largely old-regime." Martin Malia, *Alexander Herzen and the Birth of Russian Socialism* (New York: Grosset and Dunlap, 1961), p. 111.

5. F. M. Dostoevskii, *Selected Letters of Fyodor Dostoevskii*, ed. Joseph Frank and David Goldstein (New Brunswick, N.J.: Rutgers University Press, 1987), pp. 41–43.

6. "Vospominaniia V. V. Bervi (Florinskogo): Tsarstvovanie Nikolaia I.," *Golos minuvshego* 3 (1915): 138.

7. It is interesting that all four future scientists came to believe that their work was entirely compatible with religious and spiritualist views.

8. [A. N. Beketov], "Beketov, Andrei Nikolaevich, botanik," in Vengerov, *Kritiko-biograficheskii*, p. 357.

9. A. N. Beketov, *Ocherk Tiflisskoi flory s opisaniem liutikovykh, ei prinadlezhashchikh* (St. Petersburg, 1853).

10. A. N. Beketov, "Vospominaniia o Tiflise i ego Okrestnostiakh," *Vestnik Russkogo Geograficheskogo Obshchestva* 15 (1855): 65–114.

11. A. N. Beketov, *O morfologicheskikh otnosheniiakh listovykh chastei mezhdu soboiu i so steblem* (St. Petersburg, 1858).

12. A. N. Beketov, *Besedy o zemle i tvariakh, na nei zhivyshchikh* was published in St. Petersburg in 1864, 1865, 1866, 1879, 1885, 1895, 1898, and 1903.

13. Ch. Darvin, *Puteshestvie vokrug sveta na korable Bigl'* (St. Petersburg, 1865, 1871); T. G. Geksli, *O polozhenii cheloveka v riadu organicheskikh sushchestv* (St. Petersburg, 1864); A. Uolles, *Malaiskii arkhipelag* (St. Petersburg, 1872).

14. A. N. Beketov, *Kurs' botaniki dlia universitetskikh slushatelei* (St. Petersburg, 1862–1864, 1871); *Uchebnik botaniki* (St. Petersburg, 1880–1883, 1885, 1897); *Geografiia rasteniia* (St. Petersburg, 1896); "Programma po botanike dlia gimnazii," *Zhurnal Ministerstva Narodnogo Prosveshcheniia* 120, No. 12 (1863): 220–24. Among his other articles on the teaching of botany were "O prilozhenii induktivnogo metoda myshleniia k prepodavaniiu estestvennoi istorii v gimnaziiakh," *Zhurnal Ministerstva Narodnogo Prosveshcheniia* 120, No. 12 (1863): 198–224, and "Uchebnaia literatura po estestvennoi istorii v Germanii i u nas," *Zhurnal Ministerstva Narodnogo Prosveshcheniia* 134 (1867): 280–304.

15. Among them were O. V. Baranetskii (1843–1905) at Kiev University, A. N. Krasnov (1862–1914) at Khar'kov University, N. I. Kuznetsov (1864–1932) at Iur'ev University (Kuznetsov was also chair of the geobotanical section of the Main Botanical Garden of the Academy of Sciences from 1922 to 1932), I. F. Shmaulgauzen (1849–1894) at St. Petersburg and Kiev universities, and G. I. Tanfil'ev (1857–1928) at St. Petersburg and Novorossiisk universities.

16. See especially Beketov's introduction and extensive notes to his translations of one work by Grisebach: A. Grizebakh, *Rastitel'nost' zemnogo shara: Ocherk sravnitel'noi geografii rastenii*, 2 vols. (St. Petersburg, 1874, 1877).

17. A. N. Beketov, "Ob Arkhangel'skoi flore," *Trudy Sankt-Peterburgskogo Obshchestva Estestvoispytatelei* 15, No. 2 (1884): 523–616; "Fitogeograficheskii ocherk evropeiskoi Rossii," in Elize Rekliu, ed., *Zemliia i liudi* (St. Petersburg, 1884), vol. 5, vyp. 2, pp. 47–65; "Ob Ekaterinoslovskoi flore," *Botanicheskie zapiski* 1, No. 1 (1886): 1–176.

18. N. I. Kuznetsov, "Nauchnaia deiatel'nost' A. N. Beketova," *Trudy Sankt-Peterburgskogo Obshchestva Estestvoispytatelei* 33, No. 1 (1902): 241–42.

19. A. N. Beketov, "O bor'be za sushchestvovanie v organicheskom mire," *Vestnik Evropy* 10, (1873): 558–93; "Mozhno li priznavat' disgarmoniiu v prirode?" *Priroda* 1, (1876): 57–78; "Pitanie cheloveka v ego nastoiashchem i budushchem," *Vestnik Evropy* 8 (1878): 566–605; *Darvinizm s tochki zreniia obshchefizicheskikh nauk* (St. Petersburg, 1882); "Nravstvennost' i estestvoznanie," 1–67; "Doktor Froman," *Mir bozhii* 1 (1892): 47–59.

20. A. N. Beketov, *Istoricheskii ocherk dvadtsatipiati-letnei deiatel'nosti Imperatorskogo Vol'no-Ekonomicheskogo Obshchestva s 1865 do 1880 goda* (St. Petersburg, 1890).

21. For example, A. N. Beketov, "O vinograde i vine, preimushchestvenno s tsel'iu opredelit' vinogradnuiu polosu Rossii," *Vestnik Russkogo Geograficheskogo Obshchestva* 22 (1858): 1–22; "Dve publichnye lektsii ob akklimatizatsii," *Naturalist* 1 (1864): 113–15, 137–39, 189–93, 214–19; "Torfianoi ili belyi mokh i ego primenenie," *Trudy Imp. Vol'no-Ekonomicheskogo Obshchestva* 1, No. 1 (1887): 28–39; "Botanika i praktika," *Trudy Imp. Vol'no-Ekonomicheskogo Obshchestva* 4, No. 6 (1891): 286–96.

22. [Beketov], "Beketov," p. 359. Alexander Blok described his grandfather's political views with affectionate humor:

> Family head, comrade-in-arms
> Of the forties: to this day
> One of the progressives,
> He preserves the civic sacred things.
> From Nicholas' time
> He has protected enlightenment,
> But he got somewhat lost
> In the humdrum of the new movement ...
> Turgenevian tranquility
> Is part of him; he is quite
> An excellent judge of wine,
> Can appreciate delicacy in food;
> The French language and Paris
> Are perhaps closer to him than his own
> ....
> An ardent Westernizer in everything,
> At heart he is an old Russian *barin*.

Citation and translation from Mochulsky, *Aleksandr Blok*, p. 19.

23. V. O. Mikhnevich, *Nashi znakomy: Fel'etonnyi slovar' sovremennikov.* (St. Petersburg, 1885) p. 14.

24. M. A. Beketov, *Aleksandr Blok* (St. Petersburg, 1923), p. 17.

25. Students complained about Gobi's aloofness and dry pedagogical style, which they termed "*gobistika.*" Gobi was replaced by V. L. Komarov shortly after the 1917 revolution. For discussions of Beketov's stewardship as rector and his political views, see Raikov, *Russkie biologi-evoliutsionisty,* pp. 533–39, 547–48, and Shcherbakova, *Beketov,* pp. 34–46.

26. Mendeleyev shared the common Russian antipathy for Malthus. See P. T. Belov, *Filosofiia vydaiushchikhsiia russkikh estestvoispytatelei vtoroi poloviny xix—nachala xxv.* (Moscow: Mysl', 1970), pp. 317–19.

27. I. P. Borodin, "Biograficheskii ocherk zhizni A. N. Beketova," *Trudy Sankt-Peterburgskogo Obshchestva Estestvoispytatelei* 33, No. 1 (1903): 231–38; V. L. Komarov, "Slovo ot Imperatorskogo Geograficheskogo Obshchestva," *Trudy Sankt-Peterburgskogo Obshchestva Estestvoispytalelei* 33, No. 1 (1903): 282–84.

28. Alexander Blok, *Selected Poems* (USSR: Progress, 1981), p. 50.

29. A. N. Beketov, "Garmoniia v prirode," *Russkii vestnik* 30, No. 11 (1860): 542.

30. A. N. Beketov, "Garmoniia v prirode," *Russkii vestnik* 30, No. 12 (1860): 198.

31. For Beketov's admiration of Cuvier, see his review of books by John James Audobon and N. A. Severtsov, "Zhizn' ptits'," *Atenei* 13 (1858): 301–3.

32. Beketov, "Garmoniia v prirode," No. 11, 535.

33. Ibid.

34. Ibid., 534.

35. Ibid., 534, 543, 558.

36. Ibid., 537.

37. Beketov, "Garmoniia v prirode," No. 12, 201.

38. Ibid.

39. Ibid., 221–22.

40. Beketov, "Garmoniia v prirode," No. 11, 541–42.

41. Ibid., 542.

42. Ibid., 542–43.

43. By this time his student Timiriazev had presented Darwin's theory to Beketov's study circle and was working on a lengthy essay about it.

44. "Est' li prichiny predpolagat', chto formy rastenii prisposobleny k svetu?" *Naturalist* 2 (1865): 264–65; cited in Shcherbakova, *Beketov,* pp. 200–201. I have translated the words *stremliat'sia* and *stremlenie* as "to tend" and "tendency."

45. Beketov, "Dve publichnye lektsii ob akklimatizatsii," 323–24.

46. A. N. Beketov, "O bor'be za sushchestvovanie v organicheskom mire," *Vestnik Evropy* 10 (1873): 558–59.

47. Ibid., 563–64.

48. Ibid., 575.

49. Ibid., 564.

50. Ibid., 566.

51. Ibid., 567.

52. Ibid., 577. This same observation provided Kessler with the basis for his theory of mutual aid. See Chapter 6.

53. Ibid.

54. Ibid., 571.

55. Ibid., 572.

56. Ibid., 582. This statement was the closest that Beketov came to endorsing the selection theory.

57. Ibid., 572–73.

58. Ibid., 585.

59. Ibid., 586.

60. Ibid., 591–92.

61. Ibid., 586.
62. Ibid., 588-89.
63. Ibid., 593.
64. Ibid., 588.
65. Ibid., 580.
66. One example of this new balance was the fate of the Jewish people and the North American Indians. While the "national type" of these defeated peoples might disappear, they themselves continued to live as a "subordinate element" among "the stronger and more gifted people" that had defeated them.
67. Beketov, "O bor'be," 575.
68. Ibid., 564.
69. A. N. Beketov, "Mozhno li priznavat' dizgarmoniiu v prirode?" *Priroda*, 1876, 1, 57-78.
70. Mechnikov's pessimistic essay "Marital Age" appeared in *Vestnik Evropy* in January 1874 and emphasized the disharmony between the age at which humans began to experience strong sexual urges and the age at which they attained social maturity.
71. Beketov, "Mozhno li priznavat' dizgarmoniiu v prirode?" 61.
72. Ibid., 72.
73. Ibid., 74. Beketov later abandoned this view and adopted Darwin's argument that the incomplete fit between organism and environment was a telling argument against teleological views.
74. Ibid., 76.
75. Ibid., 77.
76. Ibid., 78.
77. Ibid., 77.
78. Beketov, *Uchebnik botaniki*, p. 6.
79. Ibid., pp. 4-6.
80. Ibid., pp. 513-14.
81. Ibid., p. 501.
82. Ibid., p. 506.
83. A. N. Beketov, *Darvinism s tochki zreniia obshchefizicheskikh nauk* (St. Petersburg, 1882). This essay appeared in the journal of the St. Petersburg Society of Naturalists the same year.
84. Beketov, *Uchebnik botaniki*, p. 507.
85. Ibid., p. 508. On page 509 Beketov criticized Malthus's insistence on "the fatal necessity of preventing the poor from multiplying."
86. Ibid., p. 509.
87. In her review of material in the Beketov archive at the Institute of Russian Literature of the USSR Academy of Sciences, Shcherbakova lists several undated manuscripts on natural science, philosophy, and political questions, as well as fragments of Beketov's manuscript "Nauka i nravstvennost'," written in the years 1885-1889. Shcherbakova, *Beketov*, p. 249.
88. A. N. Beketov, "Nravstvennost' i estestvoznanie," 1-67; *Geografiia rastenii: Ocherk ucheniia o rasprostraneii i raspredelenii rastitel'nosti na zemnoi poverkhnosti* (St. Petersburg, 1896). Beketov repeated the central arguments of "Morality and Natural Science," sometimes verbatim, in *The Geography of Plants*.
89. Beketov, *Geografiia rastenii*, p. 14.
90. Ibid., pp. 14-15. The same argument, almost verbatim, is found in "Nravstvennost' i estestvoznanie," 53. Beketov noted that he first enunciated this point of view in *Textbook of Botany* and in his poorly circulated article entitled "Darwinism from the Perspective of the General Physical Sciences" in 1882.
91. Beketov, *Geografiia rastenii*, p. 16; also, "Nravstvennost' i estestvoznanie," 55.
92. Beketov, *Geografiia rastenii*, p. 15. The same passage appears in "Nravstvennost i estestvoznanie," 53-54.
93. The fact that Beketov ignored indirect interspecific competition has been noted by Ia. M. Gall, *Bor'ba za sushchestvovanie kak faktor evoliutsii* (Leningrad: Nauka, 1973), p. 20.
94. Archival fragment entitled "Nauka i Nravstvennost'" in the Arkhiv Instituta Mirovoi Literatury of the Academy of Sciences, USSR, cited in Shcherbakova, *Beketov*, p. 207. The last sen-

tence of this passage probably referred to Beketov's conviction that mutual aid played an important role in the plant and animal kingdoms.

95. Beketov, "Nravstvennost' i estestvoznanie," 60.
96. Ibid.
97. Beketov, *Geografiia rastenii*, p. 16, and "Nravstvennost' i estestvoznanie," 55.
98. Beketov, *Geografiia rastenii*, p. 16.
99. Ibid., p. 17.
100. Ibid., pp. 17–18.
101. Ibid., p. 19. In "Morality and Natural Science" Beketov used the same illustration of the two horses, also mentioning Kessler's "law of mutual aid." See "Nravstvennost' i estestvoznanie," 59. The reference to Kessler was dropped in *The Geography of Plants*.
102. [Beketov], "Beketov," p. 360.
103. Beketov, *Geografiia rastenii*, p. 4.
104. Ibid., p. 27.
105. Ibid., pp. 6–7, on experimental morphology and p. 10 on the branching tree of evolution.
106. Beketov, "Doktor Froman," 50.
107. Ibid., p. 55.
108. Ibid., pp. 56–57.
109. [Beketov], "Beketov," pp. 360–61.
110. K. A. Timiriazev, *Sochineniia* (Moscow, 1939), vol. 5, p. 108.
111. V. L. Komarov, "Slovo ot Imperatorskogo Geograficheskogo Obshchestva," 283.

Chapter 4

1. K. A. Timiriazev, *Sochinenie* (Moscow, 1939), vol. 5, p. 33.
2. This is Alexander Vucinich's observation in *Science in Russian Culture, 1861–1917* (Stanford, Calif.: Stanford University Press, 1970), p. 288. Korzhinskii's work was mentioned by three speakers: August Weismann, Hugo de Vries, and George Klebs. V. O. Kovalevskii's classical monograph on the history of the horse was mentioned by one speaker, as was P. A. Kropotkin's theory of mutual aid (which was treated as a theory about humans, not other animals). See A. C. Seward, ed., *Darwin and Modern Science* (Cambridge: Cambridge University Press, 1909), pp. 24, 73, 191–92, 245, 459, 473.
3. G. D. Berdyshev and V. N. Siplivinskii, *Pervyi sibirskii professor botaniki Korzhinskii* (Novosibirsk, 1961), p. 5.
4. As V. N. Sukachev later put it, "Darwin's notion of the struggle for existence allowed him to understand the phenomena connected with the dynamics of plant cover," and became a "guiding thread" in Korzhinskii's early botanical works. Cited in Berdyshev and Siplivinskii, *Korzhinskii*, p. 8. On Korzhinskii's analysis of the struggle for existence and the dynamics of plant cover, see G. I. Dokhman, *Istoriia geobotaniki v Rossii* (Moscow, 1973), pp. 94–100; M. V. Markov, "Darvinizm i Geobotanika," *Uchenye zapiski Kazanskogo Gosudarstvennogo Universiteta* 101, No. 1 (1941): 165–81; Ia. M. Gall, *Bor'ba za sushchestvovanie kak faktor evoliutsii* (Leningrad: Nauka, 1973), pp. 36–37; and K. M. Zavadskii, *Razvitie evoliutsionnoi teorii posle Darvina* (Leningrad: Nauka, 1973), p. 164.
5. Letter from Korzhinskii to Krylov, cited in Berdyshev and Siplivinskii, *Korzhinskii*, pp. 12–14.
6. Letter from Korzhinskii to Krylov, cited in Berdyshev and Siplivinskii, *Korzhinskii*, p. 11.
7. Berdyshev and Siplivinskii, *Korzhinskii*, p. 11.
8. Ibid.
9. For the comment of someone who viewed Korzhinskii sympathetically, see A. Ia. Gordiagin, "Korzhinskii, Sergei Ivanovich," in N. P. Zagoskin, ed., *Biograficheskii slovar' professorov i prepodavatelei Imperatorskogo Kazanskogo Universiteta (1804–1904)* (Kazan, 1904), vol. 1, p. 370.
10. Stanwyn G. Shetler, *The Komarov Botanical Institute: 250 Years of Russian Research* (Washington, D.C.: Smithsonian Institution Press, 1967), p. 48.
11. Iu. A. Filipchenko, *Evoliutsionnaia ideia v biologii* [1923] (Moscow: Nauka, 1977), p. 195. The other three were Danilevskii, Timiriazev, and Kropotkin.

12. Berdyshev and Siplivinskii, *Korzhinskii*, pp. 31–32.

13. For one naturalist's comment on this, see G. I. Tanfil'ev, "O sviazi mezhdu rastitel'nogo i pochvoiu, po nabliudeniiam v voronezhskoi gubernii," *Trudy Sankt-Peterburgskogo Obshchestva Estestvoispytatelei* 22 (1892): 80. This point is discussed further in Chapter 9.

14. The most complete account of these discussions is Dokhman, *Istoriia geobotaniki v Rossii*, pp. 90–127.

15. P. A. Kropotkin, "Russia," *Encyclopaedia Britannica*, 9th ed. (Cambridge, 1886), vol. 21, p. 76.

16. Kropotkin described the antesteppe of the forest region and the intermediate zone of the steppe as areas in which "the West-European climate struggles with the Asiatic, and where a struggle is being carried on between the forest and the Steppe" (p. 77). The word "struggle" was also used in the 1880s by other observers of the boundary region between forest and steppe, for example by A. N. Krasnov (1883); see Dokhman, *Istoriia geobotaniki v Rossii*, p. 100. It was usually used, however, in a highly abstract manner and as part of an analysis emphasizing the role of physical conditions.

17. Kropotkin, "Russia," p. 77.

18. Ibid.

19. Russian naturalists agreed that the steppe had always been relatively treeless, a point established by K. F. von Baer through the absence of fossilized forms of forest fauna in the geological strata of steppe regions.

20. For contemporary reviews of this literature, see S. I. Korzhinskii, "Severnaia granitsa chernozemnostepnoi oblasti vostochnoi polosy evropeiskoi Rossii v botanikogeograficheskom i pochvennom otnoshenii: I. Vvedenie. Botanikogeograficheskii ocherk Kazanskoi gubernii," *Trudy Obshchestva Estestvoispytatelei pri Kazanskom Universitete* 18, No. 5 (1888): 20–65; G. I. Tanfil'ev, *Predely lesov na iuge Rossii* (St. Petersburg, 1894), pp. 1–27.

21. This is from the summary of Beketov's address found in *Trudy Sankt-Peterburgskogo Obshchestva Estestvoispytatelei* 16, No. 1 (1885): 46–47. Beketov had earlier expressed the same opinion of the steppe question in the notes to his translation of A. Grizebakh, *Rastitel'nost' zemnogo shara, soglasno klimaticheskomu eia raspredeleniiu* (St. Petersburg, 1877), and in the notes to his translation of Elize Rekliu, *Rossiia evropeiskaia i aziatskaia*, t. I (St. Petersburg, 1884). He amplified and defended his views in "Ob Ekaterinoslavskoi flore," *Botanicheskie zapiski* 1 (1886): 1–166, and in *Geografiia rasteniia* (St. Petersburg, 1896), pp. 352–58. In this latter work he explained how animals contributed to the dearth of forests in the steppe: "Everyone who has lived for a time on the outskirts of the steppes has probably seen the struggle of the forest with the herds: a small oak, sometimes covering a rather large area and existing in the form of a low, crooked bush, is kept in this state mainly by the herds, which gnaw around its upper shoots" (p. 25).

22. See, for example, V. Ageenko, "Flora Kryma. Tom I. Botaniko-geograficheskii ocherk tavricheskogo poluostrova," *Trudy Sankt-Peterburgskogo Obshchestva Estestvoispytatelei* 21 (1891): 1–130. For Tanfil'ev see pp. 73–74 of this volume.

23. Markov, "Darvinizm i geobotanika," 172–74. Levakovskii's experiments are summarized in Dokhman, *Istoriia geobotaniki v Rossii*, pp. 92–94, and Gall, *Bor'ba*, pp. 38–40.

24. N. F. Levakovskii, "K voprosu o vytesnenii odnikh rastenii drugimi: I. Otnoshenie semian' rastenii k vlage," *Trudy Obshchestva Estestvoispytatelei pri Imperatorskom Kazanskom Universitete* 1 (1871): 37.

25. Compare Levakovskii, "K voprosu," 36–37 with Darwin, *Origin of Species*, pp. 68–69.

26. Levakovskii, "K voprosu," 37. The inspiration for even this latter thought may well have come from Darwin's comment about his own casual experiment: "On a piece of ground three feet long and two wide, dug and cleared, and where there could be no choking from other plants, I marked all the seedlings of our native weeds as they came up, and out of the 357 no less than 295 were destroyed, chiefly by slugs and insects." Darwin, *Origin of Species*, p. 67.

27. Levakovskii, "K voprosu," 50–51.

28. N. Levakovskii, "K voprosu o vytesnenii odnikh rastenii drugimi: II. Znachenie semian' i podzemnykh chastei rastenii nakhodiashchikhsiia v pochve," *Trudy Obshchestva Estestvoispytatelei pri Imperatorskom Kazanskom Universitete* 2 (1873): 19.

29. Levakovskii, "K voprosu" (1873): 20–21.

30. Ibid., 31.
31. Ibid., 19-20.
32. S. I. Korzhinskii, "Severnaia granitsa chernozemnostepnoi oblasti vostochnoi polosy evropeiskoi Rossii v botanikogeograficheskom i pochvennom otnoshenii," *Trudy Obshchestva Estestvoispytatelei pri Imperatorskom Kazanskom Universitete* 22, No. 6 (1891): 174.
33. Ibid., 144-45.
34. Korzhinskii, "Severnaia" (1888), 62-63.
35. Ibid., 78-79.
36. Ibid., 230-31.
37. Ibid., 236.
38. Ibid., 231. The degree of interconnectedness in the bushy steppe gave it a character intermediate between the steppe and forest formations.
39. Korzhinskii, "Severnaia" (1891), 172-73.
40. Ibid., 163.
41. Ibid., 145.
42. Ibid., 150.
43. Ibid., 151-52.
44. Ibid., 160.
45. Ibid., 174-75.
46. S. I. Korzhinskii, "Chto takoe zhizn'?" in *Pervyi universitet v Sibiri* (Tomsk, 1889), p. 45.
47. Ibid., pp. 56-57. One feature of this "memory" was the influence of phylogeny on ontogeny.
48. Ibid., pp. 57-58.
49. Dokhman, *Istoriia geobotanikii v Rossii*, pp. 99-100. Dokhman adds the interesting observation that the notion of a struggle between woody and grassy types of vegetation was common in forestry, while that of a determinate influence of soil upon vegetation was central to soil scientists (p. 186). For a summary of debates from an ally of Korzhinskii's, see P. A. Kostychev, "Sviaz' mezhdu pochvami i nekotorymi rastitel'nymi formatsiiami," *Botanicheskie zapiski* 3 (1890-1893): 37-60.
50. For the perception of a sympathetic commentator, see Gordiagin, "Korzhinskii," p. 370.
51. V. V. Dokuchaev, "Referat i kriticheskii razbor geobotanicheskikh issledovanii professora Korzhinskogo v basseine kamy i chasti Volgi" (1888), *Sochinenie* (Moscow-Leningrad, 1951), vol. 6, p. 209.
52. S. T. Belozorov, *Gavriil Ivanovich Tanfil'ev* (Moscow, 1951). On Tanfil'ev's scientific work see Dokhman, *Istoriia geobotaniki v Rossii*, pp. 105-9.
53. Tanfil'ev, *Predely lesov na iuge Rossii*, pp. 21-22.
54. Ibid., p. 22.
55. Korzhinskii, "Severnaia" (1888), 79-80.
56. Tanfil'ev, *Predely lesov na iuge Rossii*, pp. 23-25.
57. Ibid., p. 27.
58. See Dokhman, *Istoriia geobotanikii v Rossii*, pp. 105-9.
59. Beketov, *Geografiia rasteniia*, p. 73.
60. Ibid., p. 74. Beketov's comment that it was more accurate to speak of the effect of conditions upon specific trees than upon a forest reflected his resistance to Korzhinskii's argument that formations should be considered as a botanical unit.
61. See Chapter 3, especially pp. 53, 56-60.
62. Cited in Dokhman, *Istoriia geobotanikii v Rossii*, p. 126.
63. K. A. Timiriazev, *Sochinenie* (Moscow, 1939), vol. 5, p. 33.
64. S. I. Korzhinskii, *Flora vostoka evropeiskoi Rossii v eia sistematicheskikh i geograficheskikh otnosheniiakh* (Tomsk, 1892), p. 2.
65. Ibid., p. 5.
66. Ibid., p. 24.
67. Korzhinskii mentions this goal in *Flora Vostoka*, p. 6.
68. Ibid., p. 49.
69. Ibid., p. 50.
70. S. I. Korzhinskii, *Ocherki rastitel'nosti Turkestana* (St. Petersburg, 1896), p. 36.

71. Korzhinskii sketched his theory in "Geterogenezis i evoliutsiia (predvaritel'noe soobshchenie)," *Izvestiia Imperatorskoi Akademii Nauk* 10, No. 3 (March 1899): 255-68, which was also published in German: "Heterogenesis und Evolution," *Naturwissenschaftl. Wochenschrift* 24 (1899): 273-78. The first half of his extended essay appeared as "Geterogenezis' i evoliutsiia: K teorii proiskhozhdeniia vidov'. I.," *Zapiski Imperatorskoi Akademii Nauk po fiziko-matematicheskomu otdeleniiu* 9, No. 2 (1900): 1-94, and in German: "Heterogenesis und Evolution, *Flora oder Allgemeine Botanische Zeitung* 89 (1901): 240-363. He also discussed heterogenesis in *Ampeolografiia Kryma* (St. Petersburg, 1904), especially pp. 21-29.
72. Korzhinskii, "Geterogenezis i evoliutsiia" (1900), 2.
73. Ibid.
74. Ibid.
75. Ibid., 78.
76. Ibid., 87.
77. Korzhinskii, "Geterogenezis i evoliutsiia" (1899), 262-64.
78. Ibid., 266. Korzhinskii acknowledged that variations sometimes occurred at a species' outer geographical limits, but attributed this to the direct action of the environment.
79. Ibid., 266-67. See also "Geterogenezis i Evoliutsiia" (1900), 77-80, 87.
80. Korzhinskii, "Geterogenezis i evoliutsiia" (1899), 267.
81. S. I. Korzhinskii, "O kleistogamii vidov Campanula," *Izvestiia Imperatorskoi Akademii Nauk* 9, No. 5 (1898): 433. My thanks to Professor Linda Berg of the Department of Botany at the University of Maryland for discussing with me the nature of cleistogamy.
82. Korzhinskii, "Geterogenezis i evoliutsiia" (1899), 267.
83. Ibid., 268.
84. Ibid.
85. K. A. Timiriazev, *Sochinenie* (Moscow, 1939), vol. 7, pp. 230-31.
86. Gordiagin, "Korzhinskii," p. 370.

Chapter 5

1. I. I. Mechnikov, "Ocherk voprosa proiskhozhdeniia vidov" [1876], *Akademicheskoe sobranie sochinenii [AS]* (Moscow, 1950), vol. 4, p. 273.
2. I. I. Mechnikov, "Über die Immunität bei Infektionskrankheiten mit besonderer Berücksichtigung der Cellulartheorie," first published in *Ergebnisse der allgemeinen Pathologie und pathologischen Anatomie* (1895). Citation is from Russian translation in *AS*, vol. 7, p. 130.
3. For discussions of Mechnikov's response to Darwin's theory, see Ia. M. Gall, *Bor'ba za sushchestvovanie kak faktor evoliutsii* (Leningrad, Nauka, 1973), pp. 21-24; K. M. Zavadskii, *Razvitie evoliutsionnoi teorii posle Darvina* (Leningrad: Nauka, 1973), pp. 147-52; A. E. Gaisinovich, "I. I. Mechnikov, velikii russkii biolog," *Zhurnal obshchei biologii*, 31, No. 4 (1970): 490-99; and I. M. Poliakov, "Razrabotka osnovnykh problem darvinizma v trudakh I. I. Mechnikova," in I. I. Mechnikov, *Akademicheskoe sobranie sochinenii* (Moscow: 1950), vol. 4, pp. 409-66. For an intellectual biography written by one of his students, see A. Besredka, *Histoire d'une idée* (Paris, 1921). Besredka sees phagocytosis as the guiding idea in Mechnikov's career, but this ignores Mechnikov's two decades of zoological research before 1883. For biographical information see also Olga Metchnikoff, *Life of Elie Metchnikoff*, (Boston and New York: Houghton Mifflin Company, 1921), and Semyon Zalkind, *Ilya Mechnikov: His Life and Work* (Moscow: 1959). A definitive biography of Mechnikov remains to be written.
4. Mechnikov referred to his brothers Nikolai (p. 317) and Ivan (p. 318) in Elie Metchnikoff, *The Prolongation of Life: Optimistic Studies* [1907], English translation by P. Chalmers Mitchell (New York and London: G. P. Putnam and Sons, 1908). Although Tolstoy's literary portrait of Ivan Il'ich was most unflattering, he referred to him in his letters as "a very kind man." See N. N. Gusev, *Lev Nikolaevich Tolstoi* (Moscow: Nauka, 1970), p. 88, and Leo Tolstoy, *The Death of Ivan Ilych and Other Stories* (New York: New American Library, 1960), pp. 95-156.
5. O. Metchnikoff, *Life of Elie Metchnikoff*, p. 29.

6. Ibid., p. 48. For L. I. Mechnikov's articles criticizing Malthusianism and "Social Darwinism," see Chapter 2, p. 39.

7. I. I. Mechnikov, *Stranitsy vospominanii* (Moscow, 1946), and *Sorok let iskaniia ratsional'nogo mirovozzreniia* (Moscow, 1913).

8. O. Metchnikoff, *Life of Elie Metchnikoff*, p. 33. Shortly after its acceptance for publication, Mechnikov discovered a serious error in his work (he had mistaken degeneration for reproduction) and sought to withdraw it from the *Bulletin of the Moscow Society of Naturalists*. His memoirs and his wife's biography of him indicate that they never realized that this first work was published despite his requests to the contrary. See Zalkind, *Mechnikov*, pp. 16–17.

9. Timiriazev, *Sochinenie* (Moscow, 1939), vol. 8, pp. 162–63.

10. Mechnikov read Bronn's German edition of *Origin of Species*.

11. O. Metchnikoff, *Life of Elie Metchnikoff*, p. 41.

12. Ibid., p. 50.

13. Mechnikov's research on comparative embryology, published between 1865 and 1883, is collected in *AS*, volumes 2 and 3.

14. Mechnikov, *Sorok let iskaniia*, in *AS*, vol. 13, p. 38.

15. O. Metchnikoff, *Life of Elie Metchnikoff*, pp. 80–81.

16. He never lost interest in the Ephemeridae and used them in *The Nature of Man* (1903) to illustrate "natural death" in the animal world.

17. I. I. Mechnikov, "Issledovaniia o vnutrikletochnom pishchevarenii u bezpozvonochnykh" (1883), *AS*, vol. 6, p. 3.

18. I. I. Mechnikov, "Rasskaz o tom, kak i pochemu ia poselilsia za granitsei" (1909), in *Stranitsy vospominaniia*, pp. 77–86. After his father-in-law's death Mechnikov managed the two family estates. The relations with the peasantry on one were very tense. In 1884 a group of peasants killed a watchman, and twelve were punished with exile to Sakhalin.

19. See John F. Hutchinson, "Tsarist Russia and the Bacteriological Revolution," *Journal of the History of Medicine and Allied Sciences* 40, No. 4 (1985): 420–39, and Zalkind, *Mechnikov*, pp. 95–102.

20. Emile Duclaux speculated that this was the reason for Pasteur's invitation to Mechnikov. See his *Pasteur: The History of a Mind* (Metuchen, N.J.: Scarecrow Reprint Corporation, 1973), pp. 317–18. Interestingly, while attempting to explain the resistance of living fowl to anthrax, Pasteur had, in 1877, hypothesized a passive competition for resources in the blood between bodily cells and bacteria. Terming this "*la lutte pour la vie*," Pasteur apologized immediately for his "*langage image*." See *Oeuvres de Pasteur*, ed. Pasteur Vallery-Radot (Paris: Masson, 1939), vol. 6, p. 177. As Duclaux indicates, Pasteur was aware of the problems with his own theory and "was conscious of not having laid hold of the vital point of the mechanism of the resistance of the organism" (p. 317). Mechnikov's concept, the "idea of a biologist and of a naturalist," had never occurred to him (p. 318). Mechnikov later recalled that Pasteur informed him during their first meeting that he thought the phagocytic theory was "on the right road," and that Pasteur had "for many years been struck by the struggle between the diverse microorganisms which I have had occasion to observe." O. Metchnikoff, *Life of Elie Metchnikoff*, p. 132. Like Duclaux, Mechnikov also noted that Pasteur's interpretation of immunity reflected his chemical training and differed fundamentally from Mechnikov's biological perspective. See *AS*, vol. 7, pp. 63, 371–72. On the relationship between Pasteur and Mechnikov, see also Yvette Conry, *L'Introduction du Darwinisme en France au xix siecle* (Paris: J. Vrin, 1974), pp. 230, 330–31. In light of Pasteur's expressed concern about the relationship between Darwinism, materialism, and immorality, Pasteur's attitude toward Mechnikov and his phagocytic theory is especially intriguing, but there is little information available. My thanks to Dr. A. M. Moulin for sharing her insights into this subject and for making available her manuscripts "Pasteurisme et Darwinisme: Les débuts difficiles de l'immunologie" and *La création du système immunitaire*.

21. Elie Metchnikoff, *The Nature of Man: Optimistic Studies*, English translation by P. Chalmers Mitchell (New York and London: G. P. Putnam and Sons, 1903), and *The Prolongation of Life*. The original French editions were published in 1902 and 1907.

22. E. Metchnikoff, *The Prolongation of Life*, p. 231. For Mechnikov's reaction to the Russian

revolution of 1905 see, for example, his letters of October 6, 1905, and August 5, 1906, in I. I. Mechnikov, *Pis'ma k O. N. Mechnikovoi (1900-1914)* (Moscow: Nauka, 1970), pp. 166 and 206-7. For a discussion of Mechnikov's political views, see also D. F. Ostrianin, *I. I. Mechnikov v bor'be za materialisticheskoe mirovozzrenie* (Kiev: Vishcha Shkola, 1977), pp. 75-88.

23. E. Metchnikoff, *The Nature of Man*, p. 295.
24. E. Metchnikoff, *The Prolongation of Life*, p. 329.
25. I. I. Mechnikov, "Uspekhi nauki v izuchenii chumy i v bor'be s neiu" [1897], in *AS* (Moscow, 1953), vol. 9, pp. 223-24. This speech was also published as "Sur la peste boubonique," *Annales de l'Institut Pasteur* 11 (1897): 737-52. In it he criticized Tolstoi and Brunetiere by name.
26. E. Metchnikoff, *The Nature of Man*, p. 166.
27. Ibid., p. 3.
28. Ibid., p. 300.
29. I. I. Mechnikov, "Neskol'ko slov o sovremennoi teorii proiskhozhdeniia vidov" [1863], first published in *AS*, vol. 4, p. 20.
30. Ibid.
31. Ibid., p. 17. For Beketov's "law of harmony" regarding fertility see Chapter 3, pp. 51-52.
32. Mechnikov, "Neskol'ko slov," pp. 17-18.
33. For Mechnikov's relationship with the radical embryologist, see A. E. Gaisinovich, "Biolog-shestidesiatnik N. D. Nozhin i ego rol' v razvitii embriologii i darvinizma v Rossii," *Zhurnal obshchei biologii* 13, No. 5 (1952): 377-87.
34. Mechnikov, "Neskol'ko slov," p. 20.
35. I. I. Mechnikov, "Zadachi sovremennoi biologii," *Vestnik Evropy* 4 (1871): 761.
36. Mechnikov, "Ocherk voprosa," p. 251.
37. Mechnikov, "Bor'ba za sushchestvovanie v obshirnom smysle" [1878], in *AS* (Moscow, 1950), vol. 4, p. 329.
38. Mechnikov, "Ocherk voprosa," pp. 251-52. He made this same point in "Bor'ba," pp. 335-36. There are minor errors in this reprint of Mechnikov's essay, which was originally published in *Vestnik Evropy* 13, No. 4 (1878): 9-47. The reprinted version also omits the second part of this article, in which Mechnikov dispassionately reviewed the relationship of the struggle for existence to various economic theories. See *Vestnik Evropy* 13, No. 5 (1878): 437-83.
39. Mechnikov, "Ocherk voprosa," p. 252.
40. See, for example, "Ocherk voprosa," pp. 329-30, where Mechnikov analyzes a case of intraspecific competition caused by the threat of a predator and concludes that the major factor in the divergence of characteristics is not always competition between the most closely allied forms.
41. Mechnikov, "Ocherk voprosa," p. 254. For Darwin's argument see *Origin of Species*, pp. 105-6. The passage cited by Mechnikov follows on p. 107.
42. Mechnikov, "Ocherk voprosa," p. 254.
43. Ibid., p. 255.
44. Ibid., p. 278.
45. Ibid., pp. 272-73.
46. Mechnikov, "Bor'ba," p. 331. See also "Zadachi sovremennoi biologii," 745-50.
47. Mechnikov, "Bor'ba," p. 333.
48. Mechnikov, "Ocherk voprosa," p. 256.
49. Mechnikov, "Bor'ba," p. 16. The word *stremlenie* can be translated as "aspiration" or "attempt," which imply conscious motivation, or simply as "tendency." Most writers, for instance, used the word *stremlenie* when translating Malthus's law of the "tendency to multiply at a geometrical rate."
50. Mechnikov, "Ocherk voprosa," p. 256.
51. This passage resembles sections of "The Struggle for Existence in a General Sense" and was almost certainly written in the 1870s. It is cited in Ostrianin, *Mechnikov*, p. 77.
52. Ostrianin, *Mechnikov*, p. 77. See also Mechnikov, "Zadachi sovremennoi biologii," 751-52.
53. Among Mechnikov's publications concerning parasitology were "O parazitizme u infuzorii" (1864), "K estestvennoi istorii chervei" (1867), "Ob odnom kolonial'nom skolekse" (1868), "Obshchii ocherk parazitel'skoi zhizni" (1874), "Materialy k ucheniiu o vrednykh nasekomykh

iuga Rossii" (1880), "Issledovaniia nad orgonektidami" (1881), and numerous articles on the beet weevil. These are collected in *AS* (Moscow, 1955), vol. 1. He drew heavily upon parasites in his discussion of evolutionary theory in "Zadachi sovremennoi biologii" (1871).

54. Mechnikov, "Obshchii ocherk," p. 201.
55. Ibid.
56. Ibid.
57. Ibid., p. 226.
58. Ibid., p. 228.
59. Ibid., p. 233.
60. Ibid., p. 286.
61. Ibid., p. 287. See note 49 on the word *stremlenie*. Mechnikov almost certainly did not mean to imply conscious aspiration by his use of this word.
62. Ibid., p. 276. Mechnikov made this same point in his discussion of the evolution of human races in "Antropologiia i Darvinizm," *Vestnik Evropy* 10, No. 1 (1875): 191.
63. Mechnikov, "Bor'ba," p. 335.
64. Mechnikov hoped that his development of a successful method for combatting the beet weevil would prove sufficiently lucrative to support him and A. O. Kovalevskii. He wrote to Kovalevskii in December 1879, "If this thing succeeds . . . then, of course, neither you nor I will have anything to fear for the future. Now more than ever I dream of the success of this enterprise." I. I. Mechnikov, *Pis'ma (1863–1916 gg.)*, ed. A. E. Gaisinovich, (Moscow: Nauka, 1974), p. 110.
65. I. I. Mechnikov "Bolezni lichinok khlebnogo zhuka" [1878], *AS*, vol. 1, p. 340.
66. Ibid., p. 342.
67. See "Zamechaniia na sochinenie g. Lindemana o khlebnom zhuke" [1880], *AS*, vol. 1, pp. 275–89. This article was originally published in the journal *Sel'skoe khoziastvo i lesovodstvo* [Agriculture and Forestry].
68. Zalkind, *Mechnikov*, p. 94. For a positive evaluation of Mechnikov's proposal for combatting the beet weevil, see entymologist P. Zabarinskii's "Zalezhi lichinok khlebnogo zhuka v vostochnoi chasti odesskogo uezda v 1887 g.," *Zapiski Obshchestva Khoziaistva Iuzhnoi Rossii* 2 (1888): 31–32.
69. Cited in O. Metchnikoff, *Life of Elie Metchnikoff*, pp. 115–16.
70. He had himself observed intracellular digestion among planaria while studying in Giessen in 1862.
71. O. Metchnikoff, *Life of Elie Metchnikoff*, pp. 116–17.
72. Elie Metchnikoff, *Immunity in Infective Diseases* [1901], trans. F. G. Binnie, (New York and London: Johnson Reprint Corporation, 1968), p. 518–19.
73. Mechnikov, "Issledovaniia o vnutrikletochnom," p. 17.
74. I. I. Mechnikov, "O tselebnykh silakh organizma" [1883], *AS*, vol. 4, pp. 22–29.
75. O. Metchnikoff, *Life of Elie Metchnikoff*, p. 151.
76. L. J. Rather has noted that only in the course of the debates over the phagocytic theory did Mechnikov become "better informed with respect to developments in medical theory prior to the start of his own interest in the subject." See L. J. Rather, *Addison and the White Corpuscles: An Aspect of Nineteenth Century Biology* (Berkeley and Los Angeles: University of California Press, 1972), p. 192.
77. O. Metchnikoff, *Life of Elie Metchnikoff*, p. 126.
78. E. Metchnikoff, *Immunity in Infective Diseases*, p. 521.
79. His friend Emile Roux later commented that it was "as a zoologist" that Mechnikov had contributed so much to medical science. See "Lettre de M. E. Roux," *Annales de l'Institut Pasteur* 29 (August 1915), p. 361.
80. My discussion of the history of ideas about inflammation draws heavily on the work of L. J. Rather, particularly his *Addison and the White Corpuscles* and his article "On the Source and Development of Metaphorical Language in the History of Western Medicine," in Lloyd G. Stevenson, ed., *A Celebration of Medical History* (Baltimore and London: The Johns Hopkins University Press, 1982), pp. 135–53.
81. Rather, *Addison and the White Corpuscles*, p. 170.
82. In his popular *Principles of Medicine* (1843), C. J. B. Williams also compared inflammation

to the process "by which a thorn or other extraneous object is removed from the flesh." Cited in Rather, *Addison and the White Corpuscles*, pp. 188-89.

83. Rather, "Metaphorical Language in Western Medicine," pp. 142-45.

84. Rather, *Addison and the White Corpuscles*, p. 146.

85. Ibid., pp. 147-48. Rather also cites Jacob Henle, who, in 1840, dismissed "the old and poetic idea" that the organism did battle with disease. Rather, "Metaphorical Language in Western Medicine," p. 151.

86. Rather, *Addison and the White Corpuscles*, p. 183.

87. Virchow had passed through Messina in 1883 and had stopped by Mechnikov's lab. See O. Metchnikoff, *Life of Elie Metchnikoff*, p. 119, and E. Metchnikoff, *Immunity in Infective Diseases*, pp. 519-20. Virchow himself shortly became one of the few pathologists to endorse the phagocytic theory, doing so in the article "The Battle of Cells and Bacteria" in 1885. For a fine essay on the views of one of Mechnikov's French contemporaries, see Camille Limoges, "Natural Selection, Phagocytosis, and Preadaptation: Lucien Cuenot, 1886-1901," *Journal of the History of Medicine and Allied Sciences* 31 (1976): 176-214. There were, of course, exceptions to the consensus that Virchow described. Both Philadelphia pathologist Joseph Richardson (1869) and the bacteriologist George Sternberg (1883) suggested that the white cells, like unicellular organisms, engulfed and destroyed or disposed of foreign matter. See Gert H. Brieger, "Introduction," in E. Metchnikoff, *Immunity in Infective Diseases*, pp. xix-xxi; George M. Sternberg, "A Contribution to the Study of Bacterial Organisms Commonly Found upon Exposed Surfaces and in the Alimentary Canal of Healing Individuals," *Johns Hopkins University Biological Studies* 2 (1883): 175.

88. Rather summarizes Cohnheim's position as follows: inflammation was "a purely pathological process set in motion by physical and chemical agencies capable of producing the required 'molecular lesion.'" See Rather, *Addison and the White Corpuscles*, pp. 173-74.

89. René Dubos, "The Micro-environment of Inflammation or Metchnikoff Revisited," *Lancet* 269, 2 (1955): 1.

90. Elie Metschnikoff, "Sur la lutte des cellules de l'organisme contre l'invasion des microbes," *Annales de l'Institut Pasteur* 1 (July 1887): 321-36. Mechnikov reviewed the objections to the phagocytic theory in E. Metchnikoff, *Immunity in Infective Diseases*, pp. 521-43.

91. See "Lettre de M. Pasteur sur la rage," *Annales del'Institut Pasteur* 1, No. 1 (1887): 1-18, on p. 10.

92. Cited in A. E. Gaisinovich, ed., "Epistoliarnoe nasledie I. I. Mechnikova," in Mechnikov, *Pis'ma (1863-1916)*, (Moscow, 1974), pp. 27-28.

93. William Henry Welch, *Adaptation in Pathological Processes* [1897] (Baltimore: The Johns Hopkins University Press, 1937), pp. 54-57. Welch praised Mechnikov in 1902 for "important observations concerning the direct participation of leucocytes and other cells in the processes of infection and immunity," and stressed that this contribution was independent of the status of the phagocytic theory itself. He considered the humoral orientation "in some respects more important." William Welch, "The Huxley Lecture on Recent Studies of Immunity, with special reference to their bearing on Pathology" (1902), in William Henry Welch, *Papers and Addresses* (Baltimore: The Johns Hopkins University Press, 1920), vol. 2, pp. 361-62.

94. Historians have yet to provide a complete account of this debate. Arthur Silverstein and A. E. Gaisinovich describe victories for both positions, but Silverstein emphasizes the "humoralist tide," particularly among younger investigators, while Gaisinovich contends that the proponents of phagocytosis more than held their own. See Arthur Silverstein, "Cellular versus Humoral Immunity: Determinants and Consequences of an Epic 19th Century Battle," *Cellular Immunology* 48 (1979): 208-21; and Gaisinovich, "Epistoliarnoe nasledie." Mechnikov's comments and activities concerning the debate on inflammation can be followed in this volume of letters and two others: I. I. Mechnikov, *Pis'ma k O. N. Mechnikovoi, 1876-99*, ed. A. E. Gaisinovich (Moscow, 1978); and I. I. Mechnikov, *Pis'ma k O. N. Mechnikovoi (1900-14)*, ed. A. E. Gaisinovich (Moscow, 1980). See also N. N. Sirotinin, "Sravnitel'no-patologicheskie issledovaniia I. I. Mechnikova o vospalenii," in *AS* (Moscow, 1954) vol. 5, pp. 283-317; Zalkind, *Mechnikov*. For a discussion of inflammation theory in the Pathological Society of London, see "A Discussion of Phagocytosis and Immunity," *British Medical Journal* 1 (1892): 373-81.

95. Elie Metchnikoff, *Lectures on the Comparative Pathology of Inflammation* [1891], trans. F. A. and E. H. Starling (New York: Dover Publications, 1968), p. 187.

96. See, for example, his discussion of opsonins and the Pfeiffer reaction in *The New Hygiene: Three Lectures on the Prevention of Infectious Diseases* (Chicago: W. T. Keener and Company, 1907), pp. 22–24.
97. E. Metchnikoff, *Immunity in Infective Diseases*, p. 519.
98. Cited in Rather, *Addison and the White Corpuscles*, p. 191.
99. Ibid., p. 191.
100. Ibid., p. 192.
101. Ibid.
102. For Mechnikov's criticism of the idealist views of the late nineteenth and early twentieth centuries, and his equation of them with religious dogma, see *AS*, vol. 11, pp. 22, 147–53.
103. Mechnikov, "Issledovaniia o Vnutrikletochnom," p. 21.
104. E. Metchnikoff, *Comparative Pathology of Inflammation*, p. 94.
105. Ibid., p. 1.
106. Ibid., p. 2.
107. Ibid., p. 3.
108. Ibid., pp. 27–28.
109. Ibid., p. 28.
110. Ibid., p. 4.
111. Ibid., p. l93.
112. Mechnikov, "Über die Immunität," p. 30.
113. E. Metchnikoff, *Comparative Pathology of Inflammation*, p. 193–94.
114. Ibid., p. 194.
115. Ibid., p. 196. This same line of reasoning enabled Mechnikov to deal with a paradox related to his theory of aging. He had concluded that the excessively large intestines in higher organisms, particularly man, provided a breeding ground for microbes that attacked the organs and tissues of the body, causing premature aging and death. Why, then, did such dangerously large intestines exist at all? Mechnikov concluded that their large size had formerly been useful for mammals in the "struggle for existence"—enabling them to feed at length, flee predators, and digest at leisure. Under the much different circumstances of contemporary human life, natural selection was undoubtedly favoring those with smaller intestines. I. I. Mechnikov, "Flora Nashego Tela," *AS*, vol. 13, pp. 191–97. This article was originally published in *Memoirs and Proceedings of the Manchester Literary and Philosophical Society* 5 (1901): 1–38.
116. To take the example offered above, overpopulation had nothing to do with the conflict between humans and the parasites that afflicted them. For Mechnikov the dynamics of the "struggle for existence" here were, rather, that interspecific conflict (the struggle between humans and infecting agents) generated intraspecific competition (among children) that favored those with advanced phagocytic properties. Mechnikov's mature opinion of Malthus is not clear from his publications after 1878. The only comment I have found is in *The Prolongation of Life* (1907), where he acknowledged that medical successes might eventually lead to an overpopulation problem, but insisted that Malthus's theory had not been confirmed "in all its details." Repeating his argument of the 1870s, he noted that human reproduction decreased to compensate for the greater numbers during peacetime. See *The Prolongation of Life*, p. 214.
117. I. I. Mechnikov, "Zakon zhizni" [1891], *AS*, vol. 13, p. 155.
118. As readers of Chapter 3 will agree, Beketov's inclusion here was problematic. A. O. Kovalevskii never, to my knowledge, discussed the factors of evolution in his published work.
119. M. A. Antonovich, *Charl'z Darvin i ego teoriia* (St. Petersburg, 1896), p. 240.
120. E. Metchnikoff, *The Nature of Man*, p. 239.
121. Mechnikov wrote "La fête au l'honneur de Darwin Cambridge," *Revue scientifique* 18 (1909): 545–48 (republished in *AS*, vol. 4, pp. 395–401); "Chestvovanie Darvina v Kembridzhe," for the newspaper *Russkoe slovo* (1909) (republished in *AS*, vol. 14, pp. 98–102); and "Darvinizm i meditsina," which first appeared in the collection *Pamiati Darvina* (Moscow, 1910) (republished in *AS*, vol. 4, pp. 402–6). The fact that Mechnikov, a Russian émigré, represented France at the Cambridge ceremony speaks volumes about the status of Darwin's theory in France.
122. Mechnikov, "Chestvovanie Darvina v Kembridzhe," p. 102.
123. Mechnikov, "Darvinizm v meditsina," p. 402. This comment should not be taken to mean that Darwin's theory was entirely correct. Even in 1871, when he was openly critical of the selection

theory, Mechnikov acknowledged that it had become "one of the fundamental dogmas of science." See "Zadachi sovremennoi biologi," 761.

124. Mechnikov, "La fête," p. 397.
125. Ibid.
126. Ibid. Note that here too intraspecific competition was generated not by Malthusian overpopulation, but by interspecific conflict.
127. Mechnikov, "Chestvovanie Darvina v Kembridzhe," p. 101. Mechnikov had earlier corresponded with de Vries regarding the cause of death among plants. See E. Metchnikoff, *The Prolongation of Life*, pp. 100–101.
128. Mechnikov, "La fête," p. 399.
129. Mechnikov, "Chestvovanie Darvina v Kembridzhe," p. 102.
130. In the English edition, Mechnikov's death dominated pages 229–75, about one-sixth, of Ol'ga's biography.
131. O. Metchnikoff, *Life of Elie Metchnikoff*, p. 255.
132. Ibid., p. 251.
133. Ibid., p. 273.
134. Ibid., p. 205. Tolstoy also mentioned these conversations in an interview with a Russian journalist. See S. P. Spiro, *Besedy s L. N. Tolstym* (Moscow, 1911), p. 35.
135. From the preface to the French edition of *Founders of Modern Medicine*, cited and translated in O. Metchnikoff, *Life of Elie Metchnikoff*, p. 248. This preface is reprinted in Elie Metchnikoff, *Trois fondateurs de la mdecine moderne* (Paris: Alcan, 1933), p. 3.

Chapter 6

1. Letter, P. A. Kropotkin to Marie Goldsmith, August 15, 1909. This copy of the original is held by the Boris Nicolaevsky Collection of the Hoover Institution on War, Revolution and Peace at Stanford University. File 6, p. 351.
2. This was cited as Kessler's credo by the students who wrote him a letter of farewell upon his departure from Kiev University to St. Petersburg University in 1862. Cited in M. N. Bogdanov, "Karl Fedorovich Kessler," *Trudy Sankt-Peterburgskogo Obshchestva Estestvoispytatelei* 12, No. 2 (1882): 337.
3. For a discussion of Severtsov and Menzbir, see Chapter 8, pp. 148–49, 155–56, 200 n.30.
4. For a discussion of Poliakov, see Chapter 7, pp. 124, 129, 142.
5. See M. Glubokovskii, "K Voprosu o darvinizme," *Vera i razum* 2 (1892): 60–65. The relationship of Beketov's belief in mutual aid to his analysis of the struggle for existence is discussed in Chapter 3, pp. 53–61. For Nozhin, see Chapter 2, pp. 31–32. For Kropotkin, see Chapter 7. For Lavrov and Mikhailovskii, see Chapter 2, pp. 35–37.
6. I. I. Mechnikov, "Neskol'ko slov o sovremennoi teorii proiskhozhdeniia vidov" [1863], *Akademicheskoe sobranie sochinenii* (Moscow, 1950), vol. 4, pp. 17–18.
7. N. D. Nozhin, "Nasha nauka i uchenye: Stat'ia V," *Knizhnyi vestnik* 6 (1866): 176.
8. Bogdanov, "Kessler," 314–15.
9. K. F. Kessler, *Estestvennaia istoriia guberniia Kievskogo uchebnogo okruga*, 5 vols. (Kiev, 1851–1856). See also V. Shimkevich, "Kessler Karl Fedorovich," *Russkii biograficheskii slovar'* (St. Petersburg, 1897), vol. 8, p. 624. For a bibliography of Kessler's work, see N. N. Banina, *K. F. Kessler i ego rol' v razvitii biologii v Rossii* (Moscow-Leningrad, 1962), pp. 134–40.
10. Banina, *Kessler*, pp. 22–23.
11. Letter to A. S. Norov, 1856, cited in Bogdanov, "Kessler," 332.
12. See, for example, K. F. Kessler, *Ryby, vodiashchiiasia i vstrechaiushchiesia v Aralo-Kaspiisko-Pontiiskoi ikhtiologicheskoi oblasti* (St. Petersburg, 1877), pp. 337–42. Citation is from p. 338.
13. Banina, *Kessler*, p. 101.
14. K. F. Kessler, "Opisanie ryb, prinadlezhashchikh k semeistvam obshchim Chernomu i Kaspiiskomu moriam," *Trudy Sankt-Peterburgskogo Obshchestva Estestvoispytatelei* 5 (1874), 191. See also Kessler, *Ryby*, p. xvii.
15. Kessler, *Ryby*, p. xvii.

16. This is from Kessler's proposal to the St. Petersburg Society of Naturalists that it organize a comprehensive study of the Crimea. Cited in V. Ageenko, "Flora Kryma. Tom I. Botaniko-geograficheskii ocherk tavricheskogo poluostrova," *Trudy Sankt-Peterburgskogo Obshchestva Estestvoispytatelei* 21 (1891): 3–4.

17. Bogdanov, "Kessler," 350.

18. K. F. Kessler, "Lektsii po estestvennoi istorii ryb," *Trudy Imperatorskogo Vol'nogo Ekonomicheskogo Obshchestva* 4, No. 4 (1863): 278.

19. He mentioned Darwin in a comment about the morphology of fish. Kessler probably read the German edition of *Origin of Species* and translated "struggle for existence" as *bor'ba iz-za sushchestvovaniia* instead of the phrase *bor'ba za sushchestvovaniia* used by Rachinskii and Timiriazev in 1864 and by almost all Russian discussants thereafter.

20. K. F. Kessler, "Nekotorye zametki otnositel'no istorii domashnikh zhivotnykh," *Trudy Imperatorskogo Vol'nogo Obshchestva* 3, No. 6 (1865): 119.

21. Kessler, *Ryby*, pp. 324–25.

22. Ibid., p. 327–29.

23. Ibid., p. 332.

24. Ibid., pp. 333–36.

25. Ibid., p. 325.

26. Cited in Banina, *Kessler*, pp. 96–97.

27. K. F. Kessler, "O zakone vzaimnoi pomoshchi," *Trudy Sankt-Peterburgskogo Obshchestva Estestvoispytatelei* 11, No. 1 (1880): 124.

28. Kessler, "O zakone vzaimnoi pomoshchi," 127.

29. Ibid., 128–29. Note that Kessler did not specify here whether mutual aid was "almost more important" than the "struggle for existence" as an ecological or an evolutionary concept.

30. Ibid., 130.

31. Ibid., 132.

32. Ibid., 132–33.

33. Ibid., 133.

34. See, for example, Kessler, "O zakone vzaimnoi pomoshchi," 133, 135.

35. Ibid., 134.

36. Ibid., 130–31.

37. Ibid., 135.

38. Ibid., 134.

39. Ibid.

40. A. F. Brandt, "Sozhitel'stvo i vzaimnaia pomoshch'," *Mir bozhii* 5, No. 5 (1896): 98.

41. Peter Kropotkin, *Mutual Aid: A Factor of Evolution* (Boston: Extending Horizon Books, 1955), p. 8. For a discussion of Kropotkin's reaction to Kessler's speech, see Chapter 7, pp. 130–31.

42. *Trudy Sankt-Peterburgskogo Obshchestva Estestvoispytatelei* 12 (1881): front page.

43. Brandt, "Sozhitel'stvo i vzaimnaia pomoshch," 98.

44. V. V. Dokuchaev, "Publichnye lektsii po pochvovedeniiu i sel'skomu khoziaistvu" [1890–1900], *Sochinenie* (Moscow-Leningrad, 1953), vol. 7, p. 277.

45. Dokuchaev, "Publichnye lektsii," p. 277.

46. See Chapter 3, p. 57.

47. Glubokovskii, "K voprosu o darvinizme," 64.

48. A. F. Brandt, *Ot materializma k spiritualizmu* (Khar'kov, 1902).

49. Among his textbooks in comparative anatomy, medical zoology, and general zoology were *Kratkii kurs sravnitel'noi anatomii pozvonochnykh* (Khar'kov, 1877), *Kratkii kurs meditsinskoi zoologii* (Khar'kov, 1893), and *Uchebnik zoologii dliia studentov* (Khar'kov, 1898). Representative of his broad interests were *Über fossile medusen* (St. Petersburg, 1871), *Über Variationsrichtungen im Thierreich* (Hamburg, 1895), and *Sexualität: Eine biologische Studie* (Dorpat, 1925).

50. A. F. Brandt, "Chislennoe ravnovesie zhivotnykh v bor'be za sushchestvovanii," *Niva* 10, No. 20 (1879): 374.

51. Ibid., 375.

52. Ibid., 411.

53. Ibid., 394.
54. A. F. Brandt, "Vrednye nasekomye," *Niva* 39 (1882): 923.
55. Ibid., 926.
56. A. F. Brandt, "Zhivotnye i rasteniia i ikh vzaimodeistvie v prirode," *Niva* 33 (1892): 739.
57. Ibid., 732.
58. Brandt, "Sozhitel'stvo i vzaimnaia pomoshch'," 1.
59. Ibid., 9.
60. Ibid., 16-17.
61. Ibid., 93-98.
62. Ibid., 98.
63. Ibid.
64. Ibid., 115.
65. Ibid., 109-14.
66. Ibid., 116.
67. M. M. Filippov, "Bor'ba i kooperatsiia v organicheskom mire," *Mysl'* 3 (1881): 352.
68. For Timiriazev's defense of Darwin see Chapter 8.
69. M. M. Filippov, "Darvinizm na russkoi pochve," *Nauchnoe obozrenie* 33 (1894): 1033.
70. M. M. Filippov, "Darvinizm," *Nauchnoe obozrenie* 34 (1894): 1065.
71. Ibid., 1064.
72. Ibid., 1065.
73. Ibid., 1071.
74. Filippov, "Darvinizm," 33, 1031.
75. Filippov, "Darvinizm," 34, 1069-70.
76. Filippov, "Darvinizm," 33, 1027. A strange comment for a philosopher who apparently had no experience in the field whatsoever.
77. M. M Filippov, "Darvinizm," *Nauchnoe obozrenie* 35 (1894): 1090.
78. M. M. Filippov, "Darvinizm," *Nauchnoe obozrenie* 32 (1894): 1001.
79. V. M. Bekhterev, "Sotsial'nyi otbor i ego biologcheskoe znachenie," *Vestnik znaniia* 12 (1912): 949.
80. Ibid., 948.
81. Ibid.
82. Ibid., 949.
83. Ibid., 955.
84. N. N. Banina and G. N. Kovan'ko, *Modest Nikobevich Bogdanov* (Leningrad, 1972); M. D. Ruzkii, "Bogdanov, Modest Nikolaevich," *Biograficheskii slovar' professorov i prepodavatelei Imperatorskogo Kazanskogo Universiteta (1804-1904)*, ed. N. P. Zagoskin (Kazan, 1904), pp. 262-67; N. P. Vagner, "Modest Nikolaevich Bogdanov (biografiia)," in M. N. Bogdanov, *Iz zhizni russkoi prirody* (St. Petersburg, 1889), pp. vii-xxiv; A. Voeikov, "M. N. Bogdanov," *Zhurnal Ministerstva Narodnogo Prosveshcheniia* 256, No. 4 (1888): 139-43.
85. See the summary of Bogdanov's speech in "Sel'skokhoziastvennoe obozrenie," *Trudy Imperatorskogo Vol'no-Ekonomicheskogo Obshchestva* 1, No. 3 (1878): 359-74. Citation is from p. 359. For his analysis of the gopher problem, see "Po voprosu o sredstvakh istrebleniia suslikov," *Trudy Imperatorskogo Vol'no-Ekonomicheskogo Obshchestva* 3, No. 4 (1873): 508-22. For a bibliography of Bogdanov's work, see Banina and Kovan'ko, *Bogdanov*, pp. 121-29.
86. See, for example, his discussion of cattle breeding in "Sel'skokhoziastvennoe obozrenie," 360-65.
87. M. N. Bogdanov, *Zoologiia pozvonochnykh. Lektsii 1883-1884*, p. 47. This lithographed edition of Bogdanov's lectures is cited in Banina and Kovan'ko, *Bogdanov*, p. 68.
88. M. N. Bogdanov, *Mirskie zakhrebetniki: Ocherki iz byta zhivotnykh, seliashchikhsiia okolo cheloveka* (St. Petersburg, 1884).
89. M. N. Bogdanov, *Iz zhizni russkoi prirody: Zoologicheskii ocherki i razskazy*, with a preface by N. P. Vagner (St. Petersburg, 1889). An amended edition, edited by P. I. Brounov and V. A. Fausek, was published under the same title in St. Petersburg in 1906.
90. Bogdanov, *Mirskie zakhrebetniki*, p. 2.
91. Ibid., p. 1.

92. Ibid., p. 2.
93. Bogdanov, *Iz zhizni russkoi prirody* (1889), p. 3.
94. Ibid., p. 10.
95. Ibid., p. 25.
96. Ibid., p. 27.
97. Ibid., p. 29.
98. Bogdanov, *Iz zhizni russkoi prirody* (1906), p. 215.
99. Ibid.
100. Ibid., p. 217.
101. Ibid., p. 218.
102. Ibid., n. p. Fausek (1861–1910) was professor at the Women's Medical Institute and the Higher Women's Courses in St. Petersburg. Among his works were *Biologicheskie izsledovaniia v zakaspiiskoi oblasti* [Biological Investigations in the Region beyond the Caspian] (St. Petersburg, 1903), a biography of Linnaeus, and *Sushchnost' Zhizni* (St. Petersburg, 1903), a collection of articles on the mechanist-vitalist controversy. He also wrote the item "The Struggle for Existence" in I. E. Andreevskii, ed., *Entsiklopedicheskii slovar'* (St. Petersburg: Brokgauz-Efron, 1891), pp. 458–60.
103. See, for example, Brandt, "Chislennoe ravnovesie," 394. For Beketov's increasing reliance on the direct action of the environment, see Chapter 3. For the relationship between Mechnikov's critique of the struggle for existence and his reliance on other mechanisms, see Chapter 5.
104. For Timiriazev's and Menzbir's comments about Kropotkin, see Chapter 8, p. 123, and Chapter 7, p. 164–65.

Chapter 7

1. P. A. Kropotkin, Letter to Marie Goldsmith, August 15, 1909. Copies of Kropotkin's letters to Goldsmith are held by the Boris Nicolaevsky Collection of the Hoover Institution on War, Revolution and Peace at Stanford University. Citation from File 6, pp. 351–52. Kropotkin added that "it is interesting that the brilliant and most honest observer, Darwin, somewhere included a two-line hint about this." I have been unable to locate it.
2. M. A. Menzbir, "Kropotkin kak biolog," *Petr Kropotkin. Sbornik statei* (Petrograd, 1922), p. 99.
3. Martin A. Miller, *Kropotkin* (Chicago and London: University of Chicago Press, 1976), p. 52. Kropotkin later recalled of this period that "the current which carried minds toward natural science was irresistible ... I soon understood that whatever one's subsequent studies might be, a thorough knowledge of the natural sciences and familiarity with their methods must lie at the foundation." See Petr Kropotkin, *Memoirs of a Revolutionist* (New York: Grove Press, 1970), p. 115. For biographical information see also S. A. Anisimov, *Puteshestviia P. A. Kropotkina v 1862–1867 godakh* (Moscow, 1952); A. Borovyi and N. Lebedev, eds., *Sbornik statei posviashchennyi pamiati P. A. Kropotkina* (St. Petersburg, 1922); James Hulse, *Revolutionists in London* (Oxford: Clarendon Press, 1970); N. K. Lebedev, *Po reke Lene i po taige vostochnoi Sibiri* (Moscow, 1929); V. A. Markin, *Petr Alekseevich Kropotkin* (Moscow, 1985); N. M. Pirumova, *Petr Alekseevich Kropotkin* (Moscow, 1972); Victor Robinson, *Comrade Kropotkin* (New York: The Altrurians, 1908); and George Woodcock, *The Anarchist Prince* (New York: Schocken Books, 1971). See also Stephen Jay Gould, "Kropotkin Was No Crackpot," *Natural History* 97, 7 (1988): 12–21.
4. *Perepiska Petra i Aleksandra Kropotkiny*, ed. N. I. Lebedev, (Moscow-Leningrad, 1932), vol. 1, p. 245, from a letter of December 1861. Shelgunov's article was based on Frederick Engels's *The Condition of the Working Class in England*.
5. Kropotkin, *Memoirs*, p. 168.
6. The letters from Siberia have been republished in P. A. Kropotkin, *Pis'ma iz vostochnoi Sibiri*, compiled by V. A. Markin and E. V. Starostin (Irkutsk, 1983).
7. Kropotkin contributed the geological section, and Poliakov the zoological, to the official *Otchet ob Olekminsko-Vitemskoi ekspeditsii*, which constituted the entire third volume of *Zapiski Russkogo Obshchestva po obshchei geografii* (1873). Poliakov knew of Kropotkin's political activ-

ities, but had no interest in them. He was completing his examinations at St. Petersburg University when he became tangentially involved in the circumstances surrounding Kropotkin's arrest. See Kropotkin, *Memoirs*, pp. 335, 338-40.

8. *Perepiska*, vol. 2, p. 194.

9. Kropotkin's contributions to natural science are analyzed in Markin, *Kropotkin*, which also contains a bibliography of his scientific work.

10. See Chapter 2, pp. 35-36. Mikhailovskii apparently made little impression on Kropotkin, however. See Miller, *Kropotkin*, p. 73.

11. P. A. Kropotkin, *Issledovaniia o lednikovom periode* (St. Petersburg, 1876).

12. For Lavrov's critique of Darwin's theory and his own thoughts about mutual aid, see Chapter 2, pp. 36-37.

13. P. A. Kropotkin, "The Aurora," *Nature* 25 (1882): 329-31, 368-72; "Geography in Russia," *Nature* 26 (1882): 211; "Metamorphic Rocks of Bergsen," *Nature* 26 (1882): 211.

14. See Miller, *Kropotkin*, pp. 133-59.

15. Letter to Marie Goldsmith, August 15, 1909, File 6, p. 349.

16. For a complete bibliography see Markin, *Kropotkin*, pp. 196-98.

17. See, for example, the following articles for the ninth edition: "Siberia" (1887, vol. 22), "Tobolsk" (1888, vol. 23), "Tomsk," (1888, vol. 23), "Transbaikalia," (1888, vol. 23),"Yakutsk," (1888, vol. 24), "Yeniseisk," (1888, vol. 24). These appeared through the year 1911. For a complete bibliography see Kropotkin, *Pis'ma iz vostochnoi Sibiri*, pp. 190-91, and Markin, *Kropotkin*, pp. 196-98.

18. P. A. Kropotkin, "The Morality of Nature," *The Nineteenth Century and After* 58 (1905): 865-83; "The Theory of Evolution and Mutual Aid," *The Nineteenth Century and After* 67 (1910): 86-107; "The Direct Action of Environment on Plants," *The Nineteenth Century and After* 68 (1910): 58-77; "The Response of the Animals to Their Environment," *The Nineteenth Century and After* 68 (1910): 856-67, 1047-59; "Inheritance of Acquired Characters: Theoretical Difficulties," *The Nineteenth Century and After* 71 (March 1912): 511-31; "Inherited Variation in Animals," *The Nineteenth Century and After* 78 (1915): 112-4-44; "The Direct Action of Environment and Evolution," *The Nineteenth Century and After* 85 (January 1919), 70-89.

19. P. A. Kropotkin, "The Dissication of Eur-Asia," *The Geographical Journal* 23 (1907): 726-41.

20. Susan Faye Cannon, *Science in Culture: The Early Victorian Period* (New York: Dawson and Science History Publications, 1978), pp. 73-110.

21. My thanks to Ernst Mayr for alerting me to this point in a personal communication.

22. For Darwin's comments on Bahia, Brazil, see Chapter 1, p. 10, and the conclusion to this volume, pp. 170-71. For Alfred Russel Wallace's comments about the coincidence between the circumstances of tropical nature and the selection theory, see the conclusion, p. 170.

23. See, for example, Kropotkin, *Pis'ma iz Vostochnoi Sibiri*, pp. 137, 145-46.

24. Kropotkin, *Memoirs*, p. 98. In a letter to Marie Goldsmith from November 1909, Kropotkin recalled that he and his brother, "already transformists," had accepted Middendorf's argument that the color of polar animals, especially that of the horses of Yakutsk, was compelling evidence "against Darwinism" and for the direct influence of the environment. Middendorf had argued that the pale coloration of domesticated animals in the North could not be attributed to natural selection. File 6, p. 362.

25. *Perepiska*, vol. 2, p. 91.

26. Ibid., p. 125.

27. Ibid., pp. 123-24.

28. Ibid., pp. 122-23.

29. Ibid., p. 117. The ellipses, apparently, are in the original letter.

30. Ibid., p. 117. For further exchanges on the origin of species, see also Alexander's letter of July 4, 1863, in vol. 2, p. 117; of February 6, 1864, in vol. 2, p. 143; and Petr's of July 19, 1863, in vol. 2, p. 123.

31. Kropotkin, *Pis'ma iz vostochnoi Sibiri*, p. 89. See also pp. 48-49, 77, 93, 152.

32. Kropotkin, *Mutual Aid*, pp. vi-viii.

33. Ibid., p. 9; see also p. viii.

34. I. S. Poliakov, "Geograficheskoe rasprostranenie zhivotnykh v iugo vostochnoi chasti Lenskago basseina," *Zapiski Imperatorskogo Russkogo Geograficheskago Obshchestva po obshchei geografii* 3 (1873): 13–15.
35. Kropotkin, *Mutual Aid*, p. 35.
36. I. S. Poliakov, *Epizody iz zhizni prirody i cheloveka iz vostochnoi Sibiri*, n. p., 1877.
37. Kropotkin, *Mutual Aid*, p. 9.
38. Ibid., p. viii.
39. Ibid., p. ix. In later years Kropotkin often invoked his Siberian experience and Poliakov's zoological observations to support the theory of mutual aid. See, for example, "The Theory of Evolution and Mutual Aid," 99–100.
40. Kropotkin, *Mutual Aid*, p. 9.
41. See Chapter 2, pp. 32–34.
42. See Chapter 2, pp. 35–36.
43. [P. A. Kropotkin], "Charles Darwin," *Le Révolté*, April 29, 1882, 1.
44. [Kropotkin], "Charles Darwin," 1.
45. See Chapter 6, pp. 110–12.
46. Note that Kropotkin changed Severtsov's observation from one that emphasized differences among falcons (some cooperated, some did not) to one that applied to falcons as a whole. He also labeled the individualism of some falcons as "brigandage" and claimed that they were "decaying throughout the world." Severtsov had said only that they were less prosperous than cooperative forms. Unlike Kropotkin, Severtsov believed that cases of mutual aid were fully compatible with Darwin's reliance on intraspecific competition. See Chapter 8, pp. 151–56.
47. Kropotkin, *Memoirs*, p. 499. Huxley's article appeared in *The Nineteenth Century* 23 (February 1888): 161–80, and was reprinted in his *Evolution and Ethics and Other Essays* (New York: D. Appleton and Company, 1902), pp. 195–236. It is also reprinted in Kropotkin, *Mutual Aid*, pp. 329–41.
48. T. H. Huxley, "The Struggle," in Kropotkin, *Mutual Aid*, p. 330.
49. Ibid., 332.
50. Ibid., p. 333.
51. Ibid.
52. Ibid.
53. Ibid., p. 334.
54. Ibid., pp. 334–37.
55. Kropotkin, *Memoirs*, p. 499.
56. Ibid.
57. Kropotkin, *Mutual Aid*, p. 60.
58. Ibid., pp. 2–3.
59. Ibid.
60. Ibid., p. 3.
61. Ibid., p. 61.
62. Ibid.
63. P. A. Kropotkin, *Fields, Factories and Workshops* (New York: G. P. Putnam's Sons, 1901), pp. 83–84.
64. Kropotkin, *Mutual Aid*, p. 68.
65. Ibid., p. 293.
66. Ibid., p. 10.
67. Ibid., p. 53.
68. Ibid., p. 74.
69. See, for example, *Mutual Aid*, p. 57.
70. Ibid., pp. 17–18.
71. Ibid., p. 57.
72. Ibid., pp. 52–53, 58. Kropotkin mentioned ants, parrots, and monkeys in this regard.
73. Ibid., p. 39.
74. Ibid., p. 52.
75. Ibid., p. 40.

76. Ibid., p. 44.
77. Ibid.
78. Kropotkin cited Moritz Wagner to this effect and integrated Wagner's argument about isolation into his view of evolution. See *Mutual Aid*, pp. 65–66. Again, the argument here is that the extermination of species does not lead to evolution.
79. Kropotkin, *Mutual Aid*, pp. 73–74.
80. Ibid., p. 295.
81. Ibid., p. 77.
82. Ibid., p. 228.
83. Ibid., p. 300.
84. Ibid., p. 58.
85. Ibid., p. 57. He also noted that mutual aid could explain the absence of certain physical traits; for instance, as a result of their cooperative life, ants did not require the individual armaments of other organisms.
86. Kropotkin, *Mutual Aid*, p. 66. Emphasis is in the original text.
87. Kropotkin, Letter to Marie Goldsmith, November 3, 1909, File 3, p. 358.
88. Shortly after the publication of the book, he wrote to Marie Goldsmith, "Mutual Aid, they say, is little by little making its way; but with the exception of two liberal newspapers, not one has said a word. Obviously they wish to be silent." This in a letter of December 12, 1902, in File 3, pp. 79–80. Within a year the first edition of *Mutual Aid* had sold out, and Kropotkin complained that the second edition was being produced with a sluggishness supposedly uncharacteristic of the English. See a letter dated December 5, 1903, and another simply dated December 1903 in File 3, pp. 85, 88.
89. See George Romanes, *Darwin and after Darwin*, 3d ed. (Chicago: Open Court, 1901), p. 269. Romanes added, however, that Kropotkin's examples of cooperation "fall under the explanatory sweep of the Darwinian theory."
90. Kropotkin, Letter to Marie Goldsmith, August 15, 1909, File 6, p. 350. In *Les théories de l'evolution* (Paris: E. Flammarion, 1909), Goldsmith and Yves Delages heartily endorsed Kropotkin's views. They cited both Lamarck and Darwin on the importance of mutual aid and commented that "the leading exponents of transformism never dreamt that their theories could be made to justify attempts at lowering man's moral status." They singled out Kropotkin and other Russian naturalists for their special contributions on this subject. Russians were attempting "to learn some lessons as to our social life, not from the life of isolated individuals, but from their observation of animal associations; this has enabled them to formulate very important and deep-searching criticisms against a too narrow application of the principle of natural selection." See *The Theories of Evolution* (New York: B. W. Huebsch, 1912), pp. 347–52. Kropotkin's comments about Russian zoology, relayed to Goldsmith in a letter of August 15, 1909, appeared almost verbatim on pages 63–64 of this work. Included was Kropotkin's claim that articles by A. F. Brandt and M. A. Menzbir on mutual aid had been inspired by his own work on that subject—a claim he sheepishly withdrew in a letter of December 25, 1909 (by which time it had already appeared in print).
91. P. Kropotkin, "Recent Science," *The Nineteenth Century* 33 (1893): 671–89; "Recent Science," *The Nineteenth Century and After* 50 (1901): 417–38. See also note 18.
92. Kropotkin, "The Theory of Evolution," 86.
93. Ibid., 87. This citation reveals, again, a "blind spot" in Kropotkin's analysis of the "struggle for existence": his failure to distinguish between indirect competition and direct combat. It also reveals his understanding of "struggle" and "cooperation" as opposing and mutually exclusive processes.
94. Kropotkin, "The Direct Action of Environment on Plants," 75.
95. Ibid., 61, 75.
96. Kropotkin, Letter to Marie Goldsmith, May 18, 1911, File 6, pp. 411–12.
97. Kropotkin, Letter to Marie Goldsmith, December 25, 1909, File 3, p. 366.
98. Kropotkin, Letter to Marie Goldsmith, March 8, 1910, File 6, pp. 380–81.
99. Kropotkin, Letter to Marie Goldsmith, August 12, 1910, File 6, p. 390.
100. Kropotkin, Letter to Marie Goldsmith, September 8, 1913, File 7, p. 462, and April 7, 1915, File 7, p. 500.

101. Kropotkin, "The Theory of Evolution," 90.
102. Ibid., 89-91.
103. Ibid., 101.
104. Kropotkin, "The Direct Action of Environment on Plants," 76.
105. Kropotkin, "The Theory of Evolution," 91.
106. Ibid.
107. Ibid., 98.
108. Ibid., 105.
109. Kropotkin, "The Direct Action of Environment and Evolution," 71-73.
110. Ibid., 89.
111. Kropotkin, *Mutual Aid*, p. 5.
112. Kropotkin, Letter to Marie Goldsmith, August 15, 1909, File 6, p. 351.
113. Ibid., p. 352.
114. Kropotkin, Letter to Marie Goldsmith, February 2, 1910, File 6, p. 378.
115. Kropotkin, Letter to Marie Goldsmith, March 8, 1910, File 6, p. 382.
116. Kropotkin, Letter to Marie Goldsmith, February 2, 1910, File 6, p. 378.
117. Kropotkin, "The Direct Action of Environment on Plants," 62. Kropotkin commented further about theories of heredity in his letter to Marie Goldsmith of April 7, 1915, in File 7, pp. 503-5.
118. Letter to Marie Goldsmith, undated, with postal stamp for August 2, 1901, File 3, p. 49.
119. He wrote to Marie Goldsmith on October 14, 1914, "The Weismannists are infuriated at me—too bad!! And I am delighted!" File 7, p. 485.
120. Kropotkin, "The Direct Action of Environment and Evolution," 75, and "The Response of the Animals to Their Environment," 864. In the first article Kropotkin argued that Weismann's mutilation of mice, intended to disprove the inheritance of acquired characteristics, "added absolutely nothing to our previous knowledge." Had Weismann bothered to read Darwin carefully, he would have understood that such mutilations were not inherited due to "embryonal regeneration"; their nontransmission, therefore, "did not affect Darwin's views upon the inheritance of variations."
121. For Kropotkin's claim that this was a misinterpretation of Lamarck, see, for instance, "The Direct Action of Environment on Plants," 77, and "The Direct Action of Environment and Evolution," 88-89. Among Kropotkin's examples of vitalist neo-Lamarckians were the Germans R. H. Francé and Adolf Wagner.
122. Kropotkin, Letter to Marie Goldsmith, December 25, 1909, File 6, p. 367.
123. Kropotkin, Letter to Marie Goldsmith, November 3, 1909, File 6, p. 358.
124. Kropotkin, Letter to Marie Goldsmith, April 7, 1915, File 7, pp. 501-3. The French in this letter is rather poor. I do not know whether the errors are Kropotkin's or the transcriber's.
125. Kropotkin, Letter to Marie Goldsmith, August 15, 1909, File 6, p. 351.
126. Ibid., pp. 351-52.
127. P. A. Kropotkin, "What to Do?" in Roger Baldwin, ed., *Kropotkin's Revolutionary Pamphlets* (New York: Dover Publications, 1971), pp. 256-57.

Chapter 8

1. K. A. Timiriazev, "U Darvina v Daune" [1909], in *Sochinenie* (Moscow, 1939), vol. 7, p. 563.
2. M. A. Menzbir, "Nikolai Alekseevich Severtsov," *Zapiski Russkogo Imperatorskogo Obshchestva po obshchei geografii* 13 (1886): ix.
3. G. P. Dement'ev, *N. A. Severtsov* (Moscow, 1948), p. 7; B. E. Raikov, *Russkii biolog-evoliutsionist Karl Frantsevich Rul'e* (Moscow-Leningrad, 1955), p. 13.
4. The best available biography of Rul'e is the Raikov volume cited in note 3.
5. N. A. Severtsov, *Periodicheskie iavleniia v zhizni zverei, ptits i gad Voronezhskoi gubernii*, ed. A. A. Grigor'ev and A. N. Formozov (Moscow, 1950), p. 26.
6. Cited in the editors' preface to Severtsov, *Periodicheskie iavleniia*, pp. 3-4.

7. See Chapter 3, note 29. Grigor'ev and Formozov claim that Severtsov's treatise was "the first detailed zoological investigation in world ecological literature," establishing Severtsov as "the founder of animal ecology." See their preface to Severtsov, *Periodicheskie iavleniia*, p. 3.

8. The evaluation of Severtsov's strengths and weaknesses by the physicomathematical faculty of Moscow University is reprinted in Dement'ev, *Severtsov*, p. 56. Raikov suggests that A. P. Bogdanov used his influence to keep the post vacant until he himself was in a position to claim it. See Rul'e, pp. 18–21.

9. Kropotkin coauthored the article entitled "Khokand" in the *Encyclopaedia Britannica*, 11th ed. (Cambridge: Cambridge University Press, 1911), vol. 15, p. 779.

10. Menzbir, "Severtsov," p. 525; A. Pankov, "Nikolai Alekseevich Severtsov, kak izsledovatel' Turkestana," *Izvestiia Turkestanskogo otdela Imperatorskogo Russkogo Geograficheskogo-Obshchestva* 11, No. 1 (1915): vvii.

11. See, for example, letters of May 1872, November 1872, and 1875 in E. P. Korovin, ed., *Russkie uchenye-issledovateli srednei Azii*. Vol. 22: *N. A. Severtsov: Sbornik dokumentov* (Tashkent, 1958), pp. 64–71, 75–82, 89–109. For Kaufmann's report on his achievements in central Asia, see *Proekt' vsepoddanneishago otcheta gen.-ad''iutata K. P. Fon-Kaufmana po grazhdanskomu upravleniiu k ustroistvu v oblastiakh Turkestanskogo general-gubernatorstva 7 noiabria 1867-25 marta 1881* (St. Petersburg, 1885).

12. The Academy of Science's charge is cited in Dement'ev, *Severtsov*, p. 23.

13. N. A. Severtsov, *Mesiats plena u kokandtsev* (St. Petersburg, 1860). This was first published in *Russkoe slovo* 70 (1859): 221–318.

14. E. F. Iunge, cited in Dement'ev, *Severtsov*, p. 58. For his comment on religion see Severtsov, *Mesiats plena u kokandtsev*, p. 66.

15. "Programma publichnykh chtenii N. A. Severtsova v S.-Peterburgskom Universitete v pol'zu bednykh studentov," reprinted in R. L. Zolotnitskaia, *N. A. Severtsov: Geograf i puteshestvennik* (Moscow, 1953), p. 195.

16. L. F. Panteleev, *Iz vospominaniia prozhlogo* (Moscow-Leningrad, 1934), p. 232.

17. Pantaleev, *Iz vospominaniia prozhlogo*, p. 232.

18. Dement'ev, *Severtsov*, p. 42.

19. For Severtsov's expression of this tension, see letters of April 1869 and November 1876 in Korovin, *Russkie uchenye-issledovateli*, pp. 55–56, 119–20. In 1878 his plan to write an evolutionary treatise entitled the "Natural History of the Turkestan Region" rather than completing a geological map of the area prompted a cut in funding. See *Russkie uchenye-issledovateli*, pp. 11–12.

20. N. A. Severtsov, *Puteshestviia po Turkestanskomu kraiu i issledovaniia gornoi strany Tiana-Shania, sovershennye po porucheniiu Russkogo Geograficheskogo Obshchestva* (St. Petersburg, 1873). For his accounts of local battles see, for example, pp. 205–14.

21. N. A. Severtsov, *Vertikal'noe i gorizontal'noe raspredelenie Turkestanskikh zhivotnykh*, published as vol. 8, no. 2, of *Izvestiia Obshchestva Liubitelei Estestvoznanii, Antropologii, i Etnografii* (Moscow, 1873). With his customary caution Severtsov referred to it, in a letter of November 1876 to Kaufmann, as a "preliminary work." See Korovin, *Russkie uchenye-issledovateli*, p. 120.

22. N. A. Severtsov, "Arkary (gornye barany)," *Priroda* 1 (1873): 144–245.

23. These materials, he noted, had been only partially exploited in his "preliminary work on Turkestan vertebrates." Letter of November 1876 to K. P. Kaufmann, in Korovin, *Russkie uchenye-issledovateli*, p. 120.

24. Letter to K. P. Kaufmann, February 1879, in Korovin, *Russkie uchenye-issledovateli*, p. 180. Severtsov mentioned similar studies by Wallace and Weismann on butterflies.

25. Cited in Raikov, *Rul'e*, p. 68.

26. These reminiscences, written in 1940, are from archival materials reproduced in Raikov, *Rul'e*, vol. 4, pp. 65–67.

27. "Vospominaniia E. F. Iunge," in Dement'ev, *Severtsov*, pp. 58–60.

28. Kropotkin, *Memoirs*, p. 228.

29. In a letter of 1880 Severtsov listed over thirty unpublished works on geography, geology, and zoogeography. See Korovin, *Russkie uchenye-issledovateli*, pp. 212–17. Dement'ev mentions two lengthy unpublished manuscripts in *Severtsov*, p. 3. See also M. Menzbir, "Severtsov, Nikolai

Alekseevich," *Entsiklopedicheskii slovar' russkogo bibliograficheskogo Instituta Granata*, 7th ed. (n.d.), vol. 41, p. 525. Severtsov informally supervised Menzbir's master's dissertation on the birds of European Russia and granted Menzbir access to his own zoological collection and unpublished material. For details of their relationship see N. Ia. Rosina, *Mikhail Aleksandrovich Menzbir* (Leningrad, 1985), pp. 12–16.

30. Menzbir became one of the most consistent defenders of the classical Darwinian approach to the struggle for existence and of the selection theory in general. See, for example, his "Poeziia i pravda eststvoznaniia," *Russkaia mysl'* 11 (1888): 124; *Darvinizm v biologii i blizkikh k nei naukakh* (Moscow, 1886), p. 54; and *Vvedenie v izuchenie zoologii i sravnitel'noi anatomii*, 2d ed. (Moscow, 1897), pp. 245–46. For Menzbir's ideas about the struggle for existence, cooperation, and sociology, see his "Obshchestva zhivotnykh," *Iuridicheskii vestnik* 1, No. 9 (1882): 27–68. My admittedly incomplete review of Menzbir's works has uncovered only one in which he mentioned the Malthus-Darwin connection—his "Pervye 65 let v istorii teorii podbora," in *Charl'z Darvin*, vol. 1, book 2, *Proiskhozhdenie vidov putem estestvennogo otbora ili sokhranenie izbrannykh porod v bor'be za zhizn'* (Moscow-Leningrad, 1926), p. 4.

31. Severtsov, *Periodicheskie iavleniia*, p. 15.
32. Ibid., p. 16.
33. Ibid., pp. 23–24.
34. Ibid., pp. 20, 37–38.
35. Ibid., p. 17.
36. Ibid., p. 18.
37. Ibid., p. 182.
38. Ibid., p. 149.
39. Ibid.
40. Ibid., pp. 179–80.
41. Ibid., p. 180. He also noted that birds migrated before temperatures reached a level that threatened their survival.
42. Ibid., p. 216. For Severtsov individual differences in moulting may well have reflected other important differences. Like many other Russian naturalists, he believed that the rate of reproduction was constantly regulated by environmental influences. The rapidity of moulting was directly proportional to a bird's fertility and inversely proportional to its long-livedness. This reflected a more general law of air-breathing animals: a species was "supported either by rapid reproduction or by the long-livedness of individuals, but never both together."
43. N. A. Severtsov, "Obez'iana-kroshka," *Vestnik estestvennykh nauk* 2, No. 26 (1855): 817–18, cited in Raikov, *Rul'e*, vol. 4, p. 34. Raikov also discusses Severtsov's "Gornye khishchiki," *Vestnik estestvennykh nauk* 5 (1854): 66–73; 9 (1854): 136–40; 11 (1854): 161–69; 33 (1854): 527–34; 34 (1854): 550–52; 36 (1854): 569–76; and "Tigr," *Vestnik estestvennykh nauk* 15–17 (1855): 19.
44. Severtsov, "Prodromus einer systematischen und zoologisch-geographischen Monographie der Katzenfamilie" (unpublished manuscript), cited in Raikov, *Rul'e*, p. 43. Severtsov alluded to this monograph in "Notice sur la classification multiseriale des félidés, et les études de zoologie générale qui s'y rattachent," *Revue et magasin de zoologie pure et appliqué*, 3d ser., 1 (1858): 148.
45. Ibid., 145–50.
46. Ibid., 195.
47. Ibid., 196.
48. N. A. Severtzow, "Notice sur la classification," *Revue et magasin*, 2d ser., 10 (Octobre 1857): 437–38.
49. Severtzov, "Notice sur la classification," 3d ser. 1, 149. One reason for his misgivings was the fact that in one and the same locale one species might be highly variable while another was relatively stable.
50. N. A. Severtzow, "Notice sur la classification," *Revue et magasin*, 2d ser., 9 (1857): 436–37.
51. N. A. Severtsov, *Ornitologiia i ornitologicheskaia geografiia evropeiskoi i aziatskoi Rossii* (St. Petersburg, 1867), cited in Dement'ev, *Severtsov*, p. 172.

52. Menzbir, "Severtsov," 527.

53. Cited in Dement'ev, *Severtsov*, p. 42. Unfortunately, Dement'ev does not give a date for this archival note, but this theme is consistent from the late 1860s to the end of Severtsov's life.

54. Korovin, *Russkie uchenye-issledovateli*, p. 213.

55. In some of his later works, Severtsov seems to rely more on direct adaptation to the environment than on natural selection. See, for example, his discussion of Capra in N. A. Severtsov,"The Mammals of Turkestan," *The Annals and Magazine of Natural History* 18 (1876): 335. This was not true, however, of others; see, for example, his "Études sur les variations d'âge des aquilines paléarctiques et leur valeur taxonomique," parts 3 and 5, *Nouveaux mémoires de la Société Impériale des Naturalistes de Moscou* 15 (1885): 84–118, and 15 (1888): 147–95.

Severtsov's biographers have analyzed his reaction to Darwin in various ways. Menzbir claimed that after Severtsov visited Darwin in 1875 he abandoned his earlier views and "became a firm proponent of [Darwin's] doctrine." See Menzbir, "Nikolai Alekseevich Severtsov," ix, and "Severtsov, Nikolai Alekseevich," p. 527. Dement'ev's analysis of this question was based largely on Menzbir's recollections. He contended that, while initially "restrained" in his attitude toward Darwin's theory, Severtsov became, by the 1870s, "one of the most visible of its proponents among zoologists" and a "convinced Darwinist." His only attempt to define the meaning of this latter phrase, however, was the comment that Severtsov thought that naturalists would uncover a complete gradation of transitions between purely individual variations and new species. See Dement'ev, *Severtsov*, pp. 42–43. Raikov pointed out that Darwin's theory elicited no "radical change" in Severtsov's views. Writing during the heyday of Lysenkoism in the USSR, Raikov contended that while Severtsov became a "fervent advocate" of Darwinism, he assigned the direct action of the environment "much more significance" than natural selection. He provided no evidence for this, nor have I found any. Raikov's comment that Severtsov's emphasis on the direct action of the environment "demonstrates, of course, his scientific perspicacity" suggests that this assessment of Severtsov's mature views was influenced by Lysenkoist Whiggism. See Raikov, *Rul'e*, pp. 56–61.

56. Severtsov, "Arkary," 245.
57. Ibid., 145.
58. Ibid.
59. Ibid., 148–49.
60. Ibid., 146–47.
61. Ibid., 167.
62. Ibid., 167–68. He made the same point in *Puteshestviia po Turkestanskomu*, 260–64.
63. Severtsov, "Arkary," 180–81.
64. Ibid., 181.
65. See Severtsov's sketches of the horns of various rams in "Arkary," 162–64. Severtsov's genus *Musimon* is now considered a species.
66. Ibid., 209–10.
67. See Severtsov's illustration of this point in "Arkary," 212.
68. Ibid., 210–11.
69. Ibid., 211.
70. Ibid., 218.

71. There was very little paleontological data bearing on this issue. Many of the areas in which the rams lived had not been explored in even a preliminary manner by zoologists.

72. Severtsov noted that the females were, therefore, less sharply differentiated than the male rams.

73. Severtsov, "Arkary," 219.

74. Ibid., 228. Severtsov cited Kropotkin's preliminary investigations of the Ice Age in his discussion of the conditions during this era. See p. 231.

75. Ibid., 241.

76. Severtsov noted that the rams, pushed from the lower pastures by the domesticated animals, themselves forced the mountain goats to higher altitudes. "Arkary," 242–43.

77. Ibid., 243.
78. Ibid., 242–43.

79. Ibid., 245.
80. N. A. Severtsov, "O zoologicheskikh (preimyshchestvenno ornitologicheskikh) oblastiakh vne-tropicheskikh chastei nashego materika," *Geograficheskie izvestiia* 13, No. 3 (1877): 148–49.
81. See the epigraph for Chapter 6 in the present volume and also Kropotkin, *Mutual Aid*, pp. 8, 228.
82. *Trudy Sankt-Peterburgskogo Obshchestva Estestvoispytatelei* 11 (1880): 119.
83. Severtsov, "Etudes sur les variations," part 3, 84. For his praise of Darwin, see especially part 3, 92.
84. K. A. Timiriazev, "Stoletnye itogi fiziologii rastenii" [1901]), in *Sochinenie*, vol. 5, p. 425.
85. K. A. Timiriazev, *Nauka i demokratiia* [1920] (Moscow: Sotsial'naia-Ekonomicheskaia Literatura, 1963), p. 11.
86. C. Timiriazeff, "The Cosmical Function of the Green Plant," *Proceedings of the Royal Society* 72 (1903): 424.
87. Timiriazev, *Sochinenie*, vol. 9, p. 46.
88. Timiriazev, "Faktory organicheskoi evoliutsii" [1890], in *Sochinenie*, vol. 5, pp. 107–8.
89. K. A. Timiriazev, "Kniga Darvina, ee kritiki i kommentatory," *Otechestvennye zapiski* 8 (1864): 880–912; 10 (1864): 550–85; 12 (1864): 859–82. It was published in St. Petersburg as a book in 1865 under the same title, and later as part 2 of *Charl'z Darvina i ego uchenie* (St. Petersburg, 1883). Between 1883 and 1936 ten more editions of this book, some with slight revisions, were published. Citations below are from *Sochinenie*, vol. 7; these passages remained unchanged from the original 1864 articles.
90. Lenin's warm response is reproduced in Timiriazev, *Sochinenie*, vol. 1, between pp. 158 and 159.
91. Timiriazev, *Sochinenie*, vol. 2, p. 13. The translation, with minor revisions, is from G. Platonov, *Kliment Arkadyevich Timiryazev* (Moscow, 1955), p. 101.
92. K. A. Timiriazev, *Solntse, zhizn', i khlorofill* [1923], in *Sochinenie*, vol. 1, p. 178.
93. K. A. Timiriazev, "La Distribution de l'energie dans le spectre solaire et la chlorophylle" [1877], in *Sochinenie*, vol. 2, p. 258.
94. K. A. Timiriazev, "L'Etat actuel de nos connaissances sur la fonction chlorphyllienne" [1884], in *Sochinenie*, vol. 1, p. 385.
95. Alfred Russel Wallace, *Darwinism* (London: Macmillan and Co., 1889), p. 302.

> The green colour of the foliage of leafy plants is due to the existence of a substance called chlorophyll, which is almost universally developed in the leaves under the action of light. It is subject to definite chemical changes during the processes of growth and of decay, and it is owing to these changes that we have the delicate tints of spring foliage, and the more varied, intense, and gorgeous hues of autumn. But these all belong to the class of intrinsic or normal colours, due to the chemical constitution of the organism; as colours they are unadaptive, and appear to have no more relation to the well being of the plants themselves than do the colours of gems and minerals.

96. K. A. Timiriazev, "Rastenie i solnechnaia energiia" [1888], in *Sochinenie*, vol. 2, pp. 257–58.
97. K. A. Timiriazev, "Stoletnye itogi fiziologii rastenii" [1901], in *Sochinenie*, vol. 5, p. 407. Translated in Platonov, *Timiryazev*, pp. 104–5.
98. References are provided in Chapter 2, note 150.
99. K. A. Timiriazev, "Osnovnye zadachi fiziologii rastenii" [1878], in *Sochinenie*, vol. 5, p. 162.
100. Timiriazev, "Faktory organicheskoi evoliutsii," pp. 139–40. Timiriazev periodically pointed to directed variations as a further source of adaptation, but always assigned the primary role to natural selection.
101. K. A. Timiriazev, *Kratkii ocherk teorii Darvina*, in *Sochinenie*, vol. 7, p. 132. This phrase, from the final, seventh edition of this work in 1919, remained unchanged from the first version, published in "Kniga Darvina, ee kritiki i kommentatory," *Otechestvennye zapiski* 10 (1864): 666.
102. Timiriazev, *Kratkii ocherk teorii Darvina*, p. 129.
103. Ibid., p. 132. He referred to the capacity of organisms to multiply in a geometrical progression on p. 137.
104. Ibid., p. 144.

105. Ibid., pp. 138-39.
106. See K. A. Timiriazev, *The Life of the Plant* [1878], trans. A. Sheremeteva (Moscow, 1958), pp. 365-67.
107. K. A. Timiriazev, "Rastenie kak istochnik sily" (1875), in *Sochinenie*, vol. 1, p. 292. The translation is A. Sheremeteva's, from *The Life of the Plant*, in which this public address was reprinted, pp. 419-20.
108. Timiriazev, "Osnovnye zadachi fiziologii rastenii" (1878), pp. 153-54. This article was reprinted without significant changes in 1895, 1904, and 1908.
109. K. A. Timiriazev, *Istoricheskii metod v biologii*, in *Sochinenie*, vol. 6, pp. 119-20. This originated as lectures in 1890-1891 and was first published in 1892-1895 in the populist journal *Russkaia mysl'*.
110. Timiriazev frequently labeled himself a "positivist" but criticized Comte for underestimating the role of theory in scientific progress. He frequently praised Comte for partially anticipating Darwin's concept of natural selection. See, for example, *The Life of the Plant*, pp. 373-74; "Charl'z Darvin," in *Sochinenie*, vol. 7, p. 220; and *Nauka i demokratiia*, pp. 230-31. For an interesting discussion of Timiriazev's philosophy of science, see N. F. Utkina, *Positivizm, antropologicheskii materializm i nauka v Rossii* (Moscow: Nauka, 1975), pp. 205-44.
111. K. A. Timiriazev, "Darvin kak obrazets uchenogo" [1878], in *Sochinenie*, vol. 7, pp. 57-58. The "windbag" in question was probably N. K. Mikhailovskii, whose articles on Darwinism in the early 1870s were, in Timiriazev's view, too weak and unconvincing to merit a response. See the introduction to the third edition of *Nasushchnye zadachi sovremennogo estestvoznaniia* [1908], in *Sochinenie*, vol. 5, p. 14. Timiriazev's revered teacher, Beketov, had voiced similar criticisms of Darwin's struggle for existence since 1873.
112. Cited in Timiriazev, "Darvin kak obrazets uchenogo," p. 56.
113. Ibid.
114. See Chapter 2, pp. 37-38, 39-43.
115. K. A. Timiriazev, "Oprovergnut li darvinizm" [1887], in *Sochinenie*, vol. 7, pp. 322-24.
116. See Chapter 2, pp. 37-38.
117. See the introduction to the third edition of *Nasushchnye zadachi sovremennogo estestvoznaniia* [1908], in *Sochinenie*, vol. 5, p. 14. Here Timiriazev also observed that the use of Darwin's theory by Haeckel and the social democratic press in Germany demonstrated that he, not Chernyshevskii, had been correct about the political implications of the selection theory.
118. Timiriazev, *Nasushchnye zadachi sovremennogo estestvoznania*, p. 31.
119. K. A. Timiriazev, "Znachenie perevorota, proizvedennogo v sovremennom estestvoznanii Darvinom" [1896], in *Sochinenie*, vol. 7, p. 252.
120. Timiriazev, "Stoletnye itogi fiziologii rastenii," p. 425.
121. K. A. Timiriazev, "Ot dela k slovu" [1907], in *Sochinenie*, vol. 5, p. 359.
122. Timiriazev, *Istoricheskii metod v biologii*, pp. 116-17.
123. K. A. Timiriazev, "Nauka i zemledelets" [1905], in *Sochinenie*, vol. 3, p. 31.
124. Timiriazev, *Istoricheskii metod v biologii*, p. 118.
125. Ibid., pp. 118-19.
126. K. A. Timiriazev, "Pervyi iubilei darvinizma" [1908], in *Sochinenie*, vol. 7, p. 575. For similar sentiments see "Stoletnye itogi fiziologii rastenii," pp. 424-25; "Ot dela k slovu," pp. 360-63; and "Nauka i vseobshchii mir" [1912] in *Nauka i demokratiia*, pp. 330-31.
127. Timiriazev, "Stoletnye itogi fiziologii rastenii," p. 425.
128. As A. E. Gaissinovitch has pointed out, Timiriazev's opposition to "Mendelism" did not imply a consistently negative appraisal of Mendel's contribution itself. See "Contradictory Appraisals by K. A. Timiriazev of Mendelian Principles and Its Subsequent Perception," *History and Philosophy of the Life Sciences* 7 (1985): 257-86.
129. Linda Gerstein is correct, I think, to assign priority here to N. N. Strakhov's "Bad Signs" (1862). See Chapter 2, p. 40.
130. Timiriazev, "Ot dela k slovu," p. 118. For a more complicated version of the same metaphor, see "Istoricheskaia biologia i ekonoicheskii materializm v istorii,"in *Sochinenie*, vol. 6, p. 228. Defending Darwin against the moral reproaches of Tolstoi and Dühring, Timiriazev explained that Darwin's "struggle for existence" was primarily defensive—"for protection from

hostile elements and all unpropitious conditions of existence—it is the struggle waged by a drowning man against the waves." Again, he omitted the element of indirect competition among "drowning men."

131. K. A. Timiriazev, "Bor'ba rasteniia s zasukhoi" (1892, 1906), in *Sochinenie*, vol. 3, pp. 177–78.

132. K. A. Timiriazev, "Ch. Darvin i K. Marks" [1919], in *Nauka i demokratiia* (Moscow, 1963), pp. 453–60.

133. Timiriazev, "Stoletnye itogi fiziologii rastenii," p. 426.

134. For an account of Kovalevskii's life and work, see Daniel P. Todes, "V. O. Kovalevskii: The Genesis, Content, and Reception of His Paleontological Work," *Studies in History of Biology* 2 (1978): 99–165. According to William Montgomery, Kovalevskii exchanged sixty-seven letters with Darwin. See "Editing the Darwin Correspondence: A Quantitative Perspective," *The British Journal for the History of Science* 20, No. 1 (1987): 20. Kovalevskii's brother, evolutionary embryologist A. O. Kovalevskii, is often labeled a classical Darwinian, but he did not, to my knowledge, ever comment publicly about the factors of evolution.

135. One example is the comment by Menzbir about Kropotkin that serves as a epigraph to Chapter 7. Similarly, Antonovich included Beketov among those Russian biologists who accepted Darwin's theory. See Chapter 5, p. 101.

Conclusion

1. Charles Lamb, "Imperfect Sympathies," in Daniel Thompson, ed., *Twenty Essays of Elia* (New York: Henry Holt and Company, 1932), p. 61.

2. See Anton Chekhov's "V usad'be," in *Sochineniia* (Moscow: 1977), vol. 8, p. 333; on the struggle for existence and Martian history, see Mark B. Adams's penetrating essay "'Red Star': Another Look at Aleksandr Bogdanov," forthcoming in *Slavic Review* (Spring 1989).

3. Yvette Conry, *L'Introduction du Darwinisme en France au xix siecle* (Paris: J. Vrin, 1974); Paul Weindling, "Theories of the Cell State in Imperial Germany," in Charles Webster, ed., *Biology, Medicine and Society, 1840–1940* (Cambridge: Cambridge University Press, 1981), p. 116; Ludwig Plate, *Uber die Bedeutung der Darwinischen Selektionsprinzips* (Leipzig and Berlin: Engelmann, 1903), and *Selektionsprinzip und Probleme der Artbildung: Ein Handbuch des Darwinismus* (Leipzig and Berlin: Engelmann, 1913). In his *No Struggle for Existence: No Natural Selection* (Edinburgh: T & T Clark, 1903), George Paulin, presumably referring to the English-language literature, complained that "no one, so far as I am aware, has ever before inquired into and examined the fundamental principles of the Darwinian theory," especially Darwin's "principle of the Struggle for Existence." This, Paulin claimed, had been accepted "as an axiomatic statement requiring neither investigation nor verification" (pp. v–vi).

4. For example, the physicogeographical factors underlying Russian non-Malthusianism were, to some degree, also present in the United States. Writing of U. S. authors' lack of interest in Malthus's theory in the years prior to 1840, G. J. Cady observed that

> European writers worked in an environment similar in its essential features to that of Malthus. The American setting was entirely different. In the United States the "short" productive factor was labor; abroad it was land. In Europe, population pressed upon the means of subsistence; in America, vast resources awaited discovery. In Europe, the problem was that of economizing productive powers; in America, the problem was that of devising the best means of exploiting them.... The dominant old world population theory to some degree reflected old world conditions; that of America reflected to some degree the buoyant confidence of a young country with almost inexhaustible natural resources.

See G. J. Cady, "The Early American Reaction to the Theory of Malthus" [1931], in John Cunningham Wood, ed., *Thomas Robert Malthus: Critical Assessments* (London: Croom Helm, 1986), vol. 4, p. 21. The U. S. case would make an especially interesting comparison with Russia because of this similarity, together with the important differences between the class structure and political traditions of the two countries.

5. See Thomas F. Glick, ed., *The Comparative Reception of Darwinism* (Austin, Tex., and Lon-

don: University of Texas Press, 1972); Peter J. Bowler, "Scientific Attitudes to Darwinism in Britain and America," in David Kohn, ed., *The Darwinian Heritage* (Princeton, N.J.: Princeton University Press, 1985), pp. 641–81; Pietro Corsi and Paul J. Weindling, "Darwinism in Germany, France and Italy," in Kohn, *The Darwinian Heritage*, pp. 683–729; Peter J. Bowler, *Evolution: The History of an Idea* (Berkeley: University of California Press, 1984); Ernst Mayr, *The Growth of Biological Thought* (Cambridge, Mass., and London: Harvard University Press, 1982). For the lessons that Chinese political thinkers drew from Darwin's "struggle for existence," see James Reeve Pusey, *China and Charles Darwin* (Cambridge, Mass.: Council on East Asian Studies, 1983).

6. R. C. Stauffer, ed., *Charles Darwin's Natural Selection* (Cambridge: Cambridge University Press, 1975), p. 176; Leonard Huxley, ed., *Life and Letters of Sir Joseph Hooker* (London: Murray, 1918), vol. 2, p. 43; T. H. Huxley, *On the Origin of Species or, The Causes of the Phenomena of Organic Nature* (New York: D. Appleton and Company, 1873), p. 120. Not all British intellectuals, of course, shared the enthusiasm of socially elite evolutionists for Malthus. See, for example, John Laurent, "Science, Society and Politics in Late Nineteenth-Century England: A Further Look at Mechanics' Institutes," *Social Studies of Science* 14 (1984): 585–619, especially 598–608.

7. Lev Tolstoi, *What to Do?*, trans. Isabel Hapgood (New York: Crowell and Company, 1887), p. 172; N. Ia. Danilevskii, *Darvinizm: Kriticheskoe issledovanie* (St. Petersburg: Komarov, 1885), vol. 1, 478; A. N. Beketov, "Nravstvennost' i estestvoznanie," *Voprosy filosofii i psikhologii* 6, No. 2 (1891): 60; P. A. Kropotkin, *Fields: Factories and Workshops* (New York: G. Putnam's Sons, 1901), 83–84.

8. P. N. Tkachev, "Nauka v poezii i poeziia v nauke" [1870], *Sochinenie v dvukh tomakh* (Moscow, 1975), vol. 1, p. 398.

9. A. I. Butovskii, *Opyt o narodnom bogatstve ili o nachalakh politicheskoi ekonomii* (St. Petersburg, 1847), p. 350, and "Obshchinnoe vladenie i sobstvennost'," *Russkii vestnik* 16 (1858): 34.

10. Danilevskii, *Darvinizm*, pp. 452–53, 461; V. V. Dokuchaev, "Publichnye lektsii po pochvovedeniiu i sel'skomu khoziaistvu" [1890–1900], *Sochinenie* (Moscow-Leningrad, 1953) vol. 7, p. 277; Mechnikov manuscript cited in D. F. Ostrianin, *I. I. Mechnikov v bor'be za materialisticheskoe mirovozzrenie* (Kiev, 1977), p. 77.

11. A. Middendorf, *Puteshestvie na sever' i vostok Sibiri:* Part 2. *Sever' i vostok Sibiri v estestvenno-istoricheskom otnoshenii* (St. Petersbrug, 1869), p. 2.

12. Alfred Russel Wallace, *Natural Selection and Tropical Nature* (London: Macmillan and Co., 1891), p. 310.

13. Wallace, *Natural Selection and Tropical Nature*, p. 310.

14. Darwin, *Autobiography*, p. 80.

15. Frederick Burkhardt and Sydney Smith, eds., *The Correspondence of Charles Darwin*, vol. 1, (Cambridge and New York: Cambridge University Press, 1985) p. 303. James R. Moore has observed, "A worshipful attitude toward nature emerged and developed following Charles's solemn encounter with the Brazilian forests in 1832." See his "Darwin of Down: The Evolutionist as Squarson-Naturalist," in Kohn, *The Darwinian Heritage*, p. 447.

16. Paul H. Barrett and R. B. Freeman, eds., *The Works of Charles Darwin*, (New York: New York University Press, 1986), vol. 2, p. 10.

17. Darwin's rhetoric about the tropics echoed that of Humboldt, whose descriptions of that region he much admired. Humboldt had previously written of the "tangled web" of tropical vegetation and compared it with a hothouse. He also observed that "organic development and abundance of vitality gradually increase from the poles towards the equator" and that tropical regions were distinguished by their "variety and magnitude of vegetable forms." Alexander von Humboldt, *Views of Nature*, trans. E. C. Otte and Henry G. Bohn (London: Henry G. Bohn, 1850), pp. 217, 230–31.

18. The "struggle over the struggle for existence" is a major motif in Mark Adams's forthcoming *Soviet Darwinism*. He sketched the contours of Soviet discussions in his manuscript "Soviet Darwinism," presented to the Darwin in Russia Symposium (1983). For a detailed discussion of the debates among Soviet botanists, see Ia. M. Gall, *Bor'ba za sushchestvovanie kak factor evoliutsii* (Leningrad: Nauka, 1973), pp. 48–118.

19. See Robert M. Young, "Darwin's Metaphor: Does Nature Select?," in Robert M. Young, *Darwin's Metaphor: Nature's Place in Victorian Culture* (Cambridge: Cambridge University Press, 1985), pp. 79–125. "Mr. Darwin's Metaphors Liable to Misconception" was first published in *The Quarterly Journal of Science* in October 1868 and was republished as "Creation by Law" in Wallace, *Natural Selection and Tropical Nature*, pp. 141–66. See *More Letters of Charles Darwin* (New York and London: Johnson Reprint Company, 1972) vol. 1, p. 268.

20. Darwin, *More Letters of Charles Darwin*, vol. 1, p. 268.

21. Charles Darwin, *On the Origin of Species* (New York: D. Appleton and Company, 1876), p. 63. This phrase first appeared in the third edition, published in 1861. For an analysis of Darwin's deliberations concerning this troublesome metaphor, see Diane B. Paul, "The Selection of the 'Survival of the Fittest,'" *Journal of the History of Biology* 21 (1988): 411–24.

Name Index

Acanfora, Michele, 4
Adams, Mark, 209n.2, 210n.18
Addison, William, 95
Ageenko, V., 67
Alcott, Louisa May, 47
Alexander II, Tsar, 26–27, 85, 122, 124
Antonovich, M. A., 23, 31, 101
Arsen'ev, K. I., 25–26
Audobon, John James, 137, 140

Bacon, Francis, 32
Baer, K. E. von, 23, 41, 66
Bakunin, M., 39, 83, 167
Baratynskii, E. A., 46
Bardakh, Ia. Iu., 86
Bates, H. W., 132
Behring, Emil, 97
Beketov, A. N., 4, 23, 65, 66, 73, 74, 75, 101, 105, 106, 113, 157, 165, 167, 169, 184n.22
  biography, 45–50
  on harmony and evolution, 50–52
  on harmony and struggle, 52–61
Beketov, N. N., 46, 116
Bekhterev, V. M., 5, 105, 116, 118–119, 166
Berg, Leo S., 174n.3
Bergson, Henri, 87, 140
Bernal, J. D., 17
Bernard, Claude, 39–40, 157
Berthelot, Marcelin, 157
Bervi-Florinskii, V. V., 46
Besredka, A., 190n.3
Bibikov, P. A., 28, 38, 44, 130, 166
Bichat, Xavier, 32
Blok, A. L., 45, 49, 50, 184n.22
Bogdanov, A. P., 23
Bogdanov, M. N., 5, 105, 107, 109, 119–21, 122, 166, 167, 168
Borodin, I. P., 50, 185n.27
Borschev, I. G., 48, 145
Boussingault, Jean, 157

Brandes, Georg, 22, 177n.70
Brandt, A. F., 5, 105, 107, 113–16, 122, 142, 167, 168
Brandt, F. F., 106, 113
Brehm, A. E., 23, 137
Bronn, H., 83
Brounov, P. I., 121
Büchner, Ludwig, 83, 124, 132
Buckle, T. R., 47, 83
Bunsen, Robert, 157
Butlerov, E. A., 46, 49
Butovskii, A. I., 26, 170

Candolle, Alphonse de, 48
Chekhov, A., 166
Cherniaev, Governor-General, 145, 146
Cherynshevskii, N. G., 22, 28, 33, 37–38, 39, 44, 135, 158, 162
Cohnheim, Julius, 94, 96
Collins, Alexis, 47
Comte, Auguste, 162
Condorcet, Marie Jean, Marquis de, 14
Cope, Edward Drinker, 140
Cuvier, Georges, 51, 107

Danilevskii, N. Ia., 5, 41–43, 44, 118, 147, 160, 167, 170
Darwin, Charles Robert, 7–13, 15–16, 19, 29, 38–39, 42, 47, 55, 57, 116, 138–42, 146, 169, 175n.24
Darwin, Francis, 157
da Vinci, Leonardo, 116
Delage, Yves, 140, 202n.90
Delianov, I. D., 49
De Vries, Hugo, 4, 62, 140
Diatropov, P. N., 86
Dokuchaev, V. V., 5, 69, 73, 75, 105, 113, 166, 167, 170
Dostoevskii, Fyodor, 32, 45, 46, 47, 84

NAME INDEX

Draper, John William, 158
Duclaux, Emile, 191n.20

Ehrlich, Paul, 86
Engels, Friedrich, 36, 37, 39, 181n.105,n.126, 199n.4
Eversmann, E. A., 46

Famintsyn, A. S., 48, 49, 65, 174n.3, 182n.150
Fausek, V. A., 199n.102, 121
Fedchenko, A. P., 108
Feuerbach, Ludwig, 83
Filipchenko, Iu. A., 65, 187n.11
Filippov, M. M., 105, 116–18
Franklin, Benjamin, 15, 163

Gall, Ia. M., 4
Galton, Francis, 12
Gamaleia, N. F., 86
Gauze, G. F., 171
Gegenbauer, Karl, 113
George, Henry, 116
Glubokovskii, M., 43, 105, 113
Gmelin, I. G., 48
Gobi, Kh. Ia., 49, 185n.25
Goldsmith, Marie, 136, 138, 140, 142
Gordiagin, A. Ia., 65, 80
Gorky, Maxim, 158
Gray, Asa, 176n.47
Grisebach, August, 48
Grove, William, 83
Gruber, Howard, 10, 19, 175n.24

Haeckel, Ernst, 113
Haxthausen, August von, 22
Helmholtz, Hermann von, 116, 157
Herbert, Sandra, 18
Herzen, A. I., 29, 83, 124, 167
Hobbes, Thomas, 41, 42
Hodge, M. J. S., 18
Hofmeister, Wilhelm, 157
Hooker, Sir Joseph, 17, 139, 169, 176n.47
Humboldt, Alexander von, 124
Hunter, John, 95
Huxley, Thomas H., 16, 29, 47, 125, 131–32, 169

Kant, I., 116
Karelin, G. S., 47
Kaufmann, K. P. von, 146
Kellogg, Vernon, 140
Kessler, K. F., 5, 23, 48, 57, 104, 115, 130, 131, 142, 147, 155, 166, 167
  biography, 105–9
  and mutual aid, 109–12

Kholodkovskii, N. A., 174n.3
Kirkhoff, Gustav, 157
Knowles, James, 132
Koch, Robert, 86, 96
Kohn, David, 18
Komarov, V. L., 48, 61
Korzhinskii, S. I., 4, 102, 166, 167, 168
  biography, 63–65
  and heterogenesis, 75–81
  on steppe-forest question, 69–72
  and vitalism, 72–75
Kossovich, P. A., 158
Kostychev, P. A., 66
Kovalevskii, A. O., 22, 23, 31, 84, 101, 167
Kovalevskii, V. O., 5, 22, 23, 165, 167
Krasheninnikov, F. N., 158
Krasnov, A. N., 66
Kropotkin, A., 124, 127
Kropotkin P. A., 4, 5, 65, 104, 105, 147, 155, 164, 166, 167, 168, 169
  biography, 124–26
  on Darwin and non-Malthusian evolutionism, 138–42
  and mutual aid, 130–38
  in Siberia, 126–30
Krylov, P. N., 63
Kutorga, S. S., 106, 107, 157
Kuznetsov, N. I., 75

Lamarck, Jean Baptiste, 109, 138, 149, 150, 160
Lamb, Charles, 173
Lavrov, P. L., 35, 105, 125, 167
Lenin, V. I., 116, 158, 207n.90
Lesgaft, P. F., 116
Leuckart, Karl, 84
Levakovskii, N. F., 5, 63, 67–69
Linnaeus, 30
Lyell, Charles, 15, 18, 23, 29
Lysenko, T. D., 171

Maack, Richard, 124
Mach, Ernst, 140
Malthus, Thomas Robert, 13–19, 24–25, 54, 55, 57, 130
Manier, Edward, 15, 20
Marx, Karl, 116, 164
Mayr, Ernst, 18
Mechnikov, I. I., 5, 22, 23, 31, 39, 44, 55, 81, 105, 165, 166, 167, 168, 170
  biography, 83–87
  death of, 102–3
  essays on selection theory, 87–91
  on parasites, 91–93
  on phagocytic theory and Darwinism, 93–102
Mechnikov, Ivan I., 83, 102
Mechnikov, L. I., 39, 83, 167

NAME INDEX

Mechnikov, O., 102
Mendeleyev, D. I., 45, 49, 116, 157
Menzbir, M. A., 5, 104, 121–22, 142, 144, 148, 151, 165
Meyer, Robert, 159
Middendorf, A. F., 68, 106, 144, 170, 200n.24
Mikhailovskii, N. K., 35–36, 44, 105, 125, 130, 166, 167
Mikhlukho-Maklai, N. N., 125
Miliutin, V. A., 26
Mill, John Stuart, 28
Moleschott, Jacques, 83, 124
Morgan, C. Lloyd, 136
Müller, Fritz, 84

Newton, I., 116
Nietzsche, Friedrich, 115, 168
Norov, A. S., 107
Nozhin, N. D., 5, 31, 44, 105, 166, 167
Nuttall, George, 97

Odoevskii, V. F., 27
Ospovat, Dov, 18
Ovsiannikov, F. V., 113

Palladin, V. I., 158
Pallas, P. S., 48, 137
Pascal, Blaise, 116
Pasteur, Louis, 86, 191n.20, 194n.91
Patkanov, K. M., 71
Pavlov, I. P., 22, 23
Peter the Great, 25–26
Petrashevskii, M. V., 46
Pfeiffer, William, 158
Pisarev, D. I., 22, 23, 28, 30, 32, 44
Plate, Ludwig, 140, 168
Plekhanov, G. V., 39, 44, 116, 167
Pobedonostev, K. P., 43, 166
Poliakov, I. S., 5, 129, 167
Poltoratskaia, S. A., 146
Prianishnikov, D. N., 158
Proudhon, Pierre-Paul, 124
Przheval'skii, N. M., 108, 125

Rachinskii, S. A., 23, 39, 146
Reclus, Jean-Jacques Elisee, 125
Reymond, Du Bois, 87
Roentgen, Wilhelm C., 116
Rogers, James Allen, 4
Rogovich, A. S., 106
Romanes, George, 136, 202n.89
Rousseau, Jean Jacques, 132
Roux, Emile, 193n.79
Royer, Clemence, 40
Rul'e, K. F., 23, 107, 127, 144, 148, 149
Rumiantsev, N., 43

Sachs, Julius von, 158
Saint-Hilare, Geoffroy, 149, 150
Schultz, Carl Heinrich, 96
Schweber, Sylvan S., 18
Scudo, Francesco, 4
Sechenov, I. M., 22, 49, 113, 167
Seidlitz, G., 88, 101
Severtsov, A. N., 146
Severtsov, N. A., 5, 23, 104, 108, 121–22, 125, 131, 142, 143, 165
  biography, 144–48
  on Darwin and the struggle for existence, 151–56
  on ecology and evolution, 1855–1858, 148–50
Shchelkov, I. P., 83
Shchurovskii, G. E., 23
Shelgunov, N. V., 29–30, 35, 124
Skvortsev, I. S., 105, 113, 166
Smith, Adam, 26, 32, 41, 42
Sokolov, N. V., 29
Spencer, Herbert, 41
Stahl, George, 95
Starr, S. Frederick, 22
Sternberg, Joseph, 194n.87
Strakhov, N. N., 40–41, 43, 44, 160, 167
Sydenham, Thomas, 95

Tanfil'ev, G. I., 67, 73, 74, 75, 167
Tarasevich, L. A., 86
Tian-Shanskii, P. P., 147
Timiriazev, K. A., 4, 5, 22, 24, 43, 44, 48, 61, 65, 75, 81, 101, 116, 122, 143, 168
  biography, 156–58
  as classical Darwinist, 159–60
  on photosynthesis, 158–59
  on struggle for existence, 160–65
Tkachev, P. N., 24, 29, 38, 44, 167, 169
Tolstoy, D. A., 108
Tolstoy, Leo, 43–44, 47, 83, 101, 102, 103, 162, 166, 167, 169
Trautfetter, R. E., 48, 106

Vagner, N. P., 46, 107, 120
Vengerov, S. A., 34
Virchow, Rudolf, 96
Vladislavlev, M. I., 39
Vogt, Karl, 29, 83, 124
Vucinich, Alexander, 4

Wagner, Moritz, 110
Wallace, Alfred Russel, 10, 15, 17, 42, 47, 55, 88, 130, 142, 159, 169, 170, 171–72, 207n.95
Weismann, August, 140–41

Welch, William Henry, 97
Wied, Prince, 137

Young, Robert M., 15, 17, 18, 171

Zabolotnyi, D. K., 86
Zaitsev, V. A., 29, 30, 32, 44
Zavodskii, K. M., 4
Zelenskii, V. V., 83
Ziegler, Ernst, 98

Subject Index

Adaptation
 Beketov on, 51, 53, 55
 Darwin-Malthus connection, 18
 Korzhinskii on, 80
 Kropotkin on, 136, 138
 and mutual aid theorists, 122
Aging, Mechnikov on, 85, 195n.115
Altruism, 116
Anarchists, 5, 123–26, 168
Antesteppe, 65–66
Anti-Malthusianism, 3–4, 26, 173–74n.2. *See also* Struggle for Existence; Malthus (name index)
Artificial selection, 19, 55–56

Balance of nature, 18, 114. *See also* Harmony of nature
Bestuzhev courses, 49
Bird migration, 106, 121, 148–49
Bolshevik revolution, 124, 126, 142, 171, 191
Botanical geography, 48, 167–68. *See also* Heterogenesis; Steppe
Botany, 4, 166
 Beketov and, 45–50
 Korzhinskii and, 63–65
 Kropotkin and, 125–26
 and population dynamics, Brandt on, 114
 steppe-forest question, 65–67, 69–72
 tropical vegetation, 210n.17
Brazil, 10, 127, 210n.15
Britain. *See* Great Britain

Capitalism, 18, 28, 34, 35, 36–37, 180n.88
Cellular immunity, 97–98
Central Asia, 145, 146, 147
Charity, Malthus on, 14
Chlorophyll, 158–60
Classical Darwinism, 143–65
 and mutual aid, 122, 123
 Severtsov and, 151–56, 165. *See also* Severtsov (name index)
 Timiriazev and, 158–65. *See also* Timiriazev (name index)
Class system
 British, 13, 19, 26, 130
 Russian, 21–22, 24, 25, 28, 168, 170
Climate. *See also* Physical conditions
 and distribution of plants, 48
 Russian, 21, 168, 169–70
Combat. *See also* Struggle for existence
 Beketov on, 58
 Darwin on, 11
 Kropotkin and, 202n.93
Communality, 77, 177n.70. *See also* Mutual Aid
Competition, 11, 33–34, 132. *See also* Conflict; Intraspecific competition and struggle; Struggle for existence
 Beketov on, 58
 combat versus, 58, 202n.93
 cooperation versus, 104–5
 Filippov on, 117–18
 Kropotkin on, 132, 134–35, 202n.93
 Levakovskii's experiments on, 67–69
 Mechnikov on, 89, 192n.40, 195n.116
 and natural selection, 10, 122
 radical thinkers and, 31
 Russian ethos, 168
 socialism and, 37
 zero-sum, 20
Conflict. *See also* Competition; Intraspecific competition and struggle; Struggle for existence
 Brandt on, 115
 cooperation versus, 104–5
 direct, 11, 132
 Filippov on, 117–18
 Kropotkin on, 132
 Mechnikov, terminology of, 84, 89, 196n.126
Conservation of energy, 158–60, 161

217

Cooperation, 36–37. *See also* Harmony of nature; Mutual aid; Sociability
  Brandt on, 115
  conflict versus, 104–5
  Darwin on, 11–12, 19
  Kropotkin on, 132, 134–35, 202n.93
  Russian attitudes, 32, 168
  Severtsov on, 155–56
Culture. *See also* Ideology
  Russian, 120, 168
  and selection theory, 19, 25, 41–42, 130, 167, 168

Darwin, evolutionary theory of. *See* Darwin (name index); Natural selection; Selection theory
Darwin, voyage of, 126–27, 170–71
Darwinism, 140, 168, 181n.104. *See also* Classical Darwinism
Darwin-Malthus connection, 13–19. *See also* Intraspecific competition and struggle; Malthusianism; Overpopulation; Population; Struggle for existence
  Mechnikov on, 88
  response to, 166–72
  Timiriazev and, 156, 160–65
Descent of Man, The, 11, 12, 39, 104
  Beketov on, 60
  conservative thinkers on, 40
  first Russian edition, 23
  Kropotkin on, 133
  Mechnikov on, 90–91
Digestion, intracellular, 93–95
Direct action of environment, 51, 109, 110, 118, 119, 122, 136–38, 150, 160, 200n.24
Disease, 16, 55. *See also* Phagocytic theory
Domestication
  Beketov on, 55–56
  Korzhinskii on, 76
  Kropotkin on, 135
  Severtsov on, 154–55

Entangled bank, 19, 171
Environment. *See* Direct action of environment; Physical conditions
Essay on the Principle of Population, An, 13–19, 130, 175n.28. *See also* Malthus (name index)
  Bibikov edition, 32–34
  British reception of, 15–29, 169, 210n.6
  Russian reception of, 25–29, 169
Evolution. *See also* Natural selection; Selection theory
  Beketov and, 48
  harmony of nature and, 10–52
  heterogenesis and, 75–81
  Mechnikov and, 98–99

  mutual aid and, 122, 136–42
    Bogdanov on, 119
    Kropotkin on, 126, 127–28, 132–38, 142
    Severtsov and, 146, 147, 148–50
Evolutionism, 23
  after assassination of Alexander II, 122
  non-Malthusian, 4, 125, 170
  radical thinkers and, 30, 31

Family life, 111, 121, 131
Famine, 13, 16
Fecundity. *See* Reproduction
Food supply, 33, 34, 42. *See also* Struggle for existence
  and fish behavior, 109
  mutual aid and, 112, 115, 121, 134, 148
Forest, 65–67, 69–72, 167
France, 169, 195n.121

Geographical isolation, and evolution, 89–90, 109–10, 139, 141
Germany, 169
Great Britain. *See also* Class system, British
  culture and selection theory, 19, 25, 38–43, 167, 168
  physical conditions, 168
  population, 177n.66
  reception of Malthus in, 15–19, 169, 210n.6

Harmony of nature, 50–61, 160. *See also* Beketov (name index)
Heterogenesis, 4, 62, 75–81. *See also* Korzhinskii (name index)
Human nature
  Mechnikov on, 84, 85, 86, 90–91
  mutual aid and
    Brandt on, 114, 116
    Kessler on, 110, 112
    Kropotkin on, 135
    Timiriazev on, 162, 163

Idealism, 140, 141
  heterogenesis as, 75
  Kropotkin on, 139–41
  Mechnikov on, 87, 195n.102
Ideology, 166, 167
  of Darwin, 13, 31
  and interpretation of Darwin, 19
  of Kropotkin, 123, 124–26, 130, 131, 142
  leftist, 29–32, 105, 166
  of Mechnikov, 86–87
  and metaphor in science, 171
  and mutual aid, 105, 123, 124–26, 130, 131, 142
  political. *See* Political ideology
  populist, 35–38

science and, 35
  of Timiriazev, 156–57
  Tsarist, 29, 157–158
Immunology, 97–98. *See also* Phagocytic theory
Individualism, 25, 36, 39, 41, 86, 122, 163
  Bogdanov on, 121
  and Russian ethos, 168
Infection. *See* Phagocytic theory
Inflammation, 95–98. *See also* Phagocytic theory
Inheritance of acquired characteristics, 126. *See also* Lamarckism
Interspecific struggle, 33–34
  Beketov on, 58, 186n.93
  Mechnikov on, 89, 99, 195n.116, 196n.126
  Severtsov on, 155
  Timiriazev on, 160
Intracellular digestion, 93–95
Intraspecific competition and struggle. *See also* Cooperation; Mutual aid; Struggle for existence
  Beketov on, 54, 57, 58, 59
  Bekhterev on, 118
  Bibikov on, 33–34
  Bogdanov on, 121
  British cultural context and, 18–19
  Darwin on, 10–11
  Filippov on, 117–18
  Kessler on, 109, 111
  Kropotkin on, 130, 132–33, 135
  Mechnikov on, 88–89, 192n.40, 195n.116, 196n.126
  mutual aid and, 104–5, 122
  Nozhin on, 31–32
  postrevolutionary ideas, 171
  Timiriazev on, 160
Isolation, geographical, 89–90, 109–10, 139, 141

Lamarckism, 37–38, 139, 141
  Beketov on, 56
  of Chernyshevskii, 37
  of Kropotkin, 126
  Severtsov and, 150
  vitalism and, 140
Liberals, 5, 28, 29, 166
Lysenkoism, 4, 171, 173–74n.2, 206n.55

Malthusianism, 168. *See also* Anti-Malthusianism; Intraspecific competition and struggle; Malthus (name index); Overpopulation; Struggle for existence
  anti- versus non-Malthusian reactions, 3–4
  Beketov and, 54–55
  of Darwin's successors, 139
  Filippov on, 117
  Kropotkin and, 131–32, 133, 139
  Pisarev on, 28–29
  Timiriazev and, 143–44, 160–65

Marxism and Marxists, 39, 116, 158
Materialism, 22, 26
Mendelism, 164, 208n.128
Metaphor, 7, 19–20, 171, 173n.1
Migration, 76, 106, 119, 121, 136, 148–49
Morality, 43, 57, 113
  Beketov on, 59, 186n.88
  Huxley argument, Kropotkin and, 131–32
  Kessler on, 112
  Korzhinskii on, 81
  Odoevskii on British, 27
  Strakhov on, 40
  Timiriazev on, 162
Morphology, 108, 118, 146, 148, 150, 154, 156
Mutation theory. *See* Heterogenesis
Mutual aid, 5, 31, 43, 54, 59, 167, 168. *See also* Kessler (name index); Kropotkin (name index)
  Bekhterev and, 118–19
  Bogdanov and, 119–21
  Brandt and, 113–16
  classical Darwinists' response, 164–65
  Filippov and, 116–18
  Glubokovskii and, 113
  Kessler and, 104–22
  Kropotkin and, 123–42
  Lavrov and, 36–37
  Mechnikov and, 87–88
  Menzbir and, 104, 205n.30
  Nozhin and, 31–32
  Poliakov and, 104, 129
  Severtsov and, 147, 155–56
  Skvortsev and, 113
  Timiriazev and, 164–65
  weak versus strong variants, 104, 105, 121–22

Naturalists, Society of, 108, 167
Natural selection, 15. *See also* Selection theory
  Darwin on, 8–9
  Filippov on, 118
  Kropotkin on, 126, 132, 136–42
  Mechnikov on, 98–102
  Timiriazev on, 160
  Wallace on, 172
Nature
  great tree of, 19
  man in, 19
  plants, role in, 159
  rationality of, 27
  and society, 35–36, 114
  struggle in natural versus social worlds, 38–39
Nature, human. *See* Human nature
Nutrition. *See* Food

On the Origin of Species by Means of Natural Selection, 28. *See also* Darwin (name index)
  Beketov and, 53, 59

On the Origin of Species by Means of Natural Selection (*continued*)
  Bogdanov and, 119
  conservative thinkers and, 39–43
  first Russian edition, 23
  Kessler and, 109
  Kropotkin and, 127, 129, 133
  and Lamarckian theories, 138–39, 141
  Mechnikov and, 84, 87–91
  metaphors in, 7–11, 19–20, 171–72, 173n.1
  radical thinkers and, 30
  Severtsov and, 146, 151
  struggle for existence described, 7–13
  Timiriazev and, 157, 160
Ornithology, 131, 168
  Bogdanov's studies, 119, 121
  Kessler's studies, 106
  Severtsov's studies, 146, 147–48, 149, 155, 156
Orthodox church, 25, 177–78n.73
Orthodox Darwinism. *See* Classical Darwinism
Overpopulation, 168. *See also* Malthusianism; Struggle for existence
  Brandt on, 115
  Butovskii on, 26
  Kropotkin on, 129, 142
  Mechnikov on, 87–91, 195n.116
  Russian attitudes toward, 21, 26, 32

Parasites
  Brandt on, 114
  Mechnikov on, 91–93, 99–100, 192–93n.53
Pathologists, 5, 22, 94, 95–98, 168. *See also* Phagocytic theory
Phagocytic theory, 5. *See also* Mechnikov (name index)
  Darwin's theory and, 98–102
  inflammation, pathologists and, 95–98
  and intracellular digestion, 93–95
  and parasitism, 91–93
Physical conditions. *See also* Environment
  and botanical geography, 48, 65–67, 69–72
  Filippov on, 117–18
  and harmony of nature, 52, 53, 54, 58
  Kropotkin on, 134, 136
  in Russia, 20–21, 168–70
  in Siberia, 128–29
  struggle with. *See* Mutual aid
  Timiriazev on, 160, 161
Phytogeography, 48, 167–68
Political ideology, 4–5, 26, 27, 28, 29–32, 166
  conservative thinkers, 39–43
  of Darwin, 12, 175n.24
  of Kropotkin, 123, 124–26, 130–31
  of Mechnikov, 86–87
  and mutual aid tradition, 105
  radical, 29–32, 38–39
  and scientists in 1860s, 22
  of Timiriazev, 156–57
Population, 24, 177n.66
  Brandt on, 114
  checks on, 13, 16, 34
  Danilevskii on, 43
  land area and, 209n.4
  Mechnikov on, 195n.116, 196n.126
  in Russia, 21
  socialism and, 26
Populism, 22, 34–38, 105, 166, 167, 168, 179n.54

Radicals, 38–39, 167
  and Malthus, 25, 29–32
  and science, 40, 107, 167
Religion, 4–5, 25, 39, 43, 87, 105, 130, 178n.73
  and evolutionary theory, 56
  Mechnikov and, 86, 87, 195n.102
  and mutual aid tradition, 105
Reproduction, 8, 195n.116
  Brandt on, 114
  harmony of nature and, 51
  Kessler on, 110, 112
  Mechnikov on, 87–88, 90
  Severtsov on, 205n.42
Revolution, Russian, 124, 126, 142, 171, 191–92n.22
Russia
  physical conditions, 4, 48, 20–21, 167, 168–70, 177n.67, 209n.4
  scientific organizations in, 22–23, 48, 61, 107, 108, 109
  social conditions, 21–22, 110, 168, 177n.67
  translations of Darwin's works, 23
  University Statute of 1884, 49, 63–64

Scientism, 29–30, 35, 40, 107, 167
Selection theory, 15, 16. *See also* Natural selection
  Beketov and, 53
  British cultural roots, 39–43
  Mechnikov on, 82, 87–91, 98–102
  mutual aid and, 121–22
  reception in Russia characterized, 23
  and socialism, 37–38
Siberia, 63, 64, 68, 124, 126–32, 142, 170
Slavophiles, 39, 41
Sociability. *See also* Cooperation; Mutual Aid
  Beketov on, 55, 59
  Darwin on, 11–12
  degrees of, 148
  Kessler on, 111
  Kropotkin on, 129, 134
  among plants, 69
  Severtsov on, 148–49
Social Darwinism, 13, 19, 32, 34, 39, 168, 181n.96
Socialism, 26, 28, 35–38, 125, 130
Social thought, Russian, 4, 125
  Bibikov edition of Malthus, 32–34
  condemnation of Malthus, 38–39
  conservative critics, 39–43

populism, natural science, and struggle for existence, 34–38
radical thinkers, 29–32
reception of Malthus 1798–1864, 25–9
Tolstoy, 43–44
Society of Naturalists, 108, 167
Speciation
  and balance, Brandt on, 114
  Kessler on, 109–10
  Korzhinskii on, 76, 78–79
  Kropotkin on, 135, 136
  Mechnikov on, 89–90
  Severtsov on, 146, 148, 150, 156
  Timiriazev on, 160
Species
  definition, 127–28, 150
  formation of. See Speciation
  struggle between. See Interspecific struggle
  struggle within. See Intraspecific competition and struggle
  variability within, 146, 148, 150, 156
Steppe, 48, 65–67, 69–72, 167
Struggle for existence. See also Interspecific struggle; Intraspecific competition and struggle; Mutual aid
  Beketov on, 52–60, 74
  Bekhterev on, 118–19
  Bibikov on, 33–34
  Bogdanov on, 119–21
  Brandt on, 113–16
  and British context, 17–19, 40–42
  Danilevskii on, 40–43
  Darwin on, 7–13
  Filippov on, 116–18
  Huxley on, 16–17
  Kessler on, 111–12
  Kropotkin on, 129–42, 202n.93
  Malthus on, 13–14
  Mechnikov on, 89–101
  as metaphor, 7–8, 19–20, 53, 133, 160–65, 168, 173n.1
  and parasitism, 91–93
  and populist thinkers, 34–38
  postrevolutionary views, 171
  and radicals, 30–32, 38–39
  and Russian context, 3–4, 166–70

Severtsov on, 151–56
Strakhov on, 40
Tanfil'ev on, 73–74
Terminology, 15, 33–34, 58, 84, 88–89, 104–5, 160–61
Timiriazev on, 160–65
Tolstoy on, 43–44
Wallace on, 171
Symbiosis, 114–15, 117

Teleology, 27, 39, 95–97, 100, 138
Translations
  of Darwin, 23, 47. See also specific works
  of Malthus, 32–34, 169, 178n.7
Tropics, 10, 142, 169, 170, 171, 210n.15, 210n.17

Underpopulation, 25–26, 169–70. See also Population
United States, land area, 290n.4

Variation
  Darwin on, 8–9
  individual, 146, 148, 150, 156
  Kessler on, 109–10, 111–12
  Korzhinskii on, 76–80
  Mechnikov on, 89
Vitalism, 72–75, 140

War, 13, 16, 19, 111
Wedges, 12, 18, 19, 169

Zoogeography
  Bogdanov, work of, 119–21
  Kessler, work of, 106
  Severtsov, work of, 143, 144–50, 204n.29
Zoologists, 5, 22, 23, 39, 166, 174n.3
  and mutual aid tradition, 105
  social responsibility of, 115
  struggle for life, observations on, 137